Lecture Notes in Mathematics

A collection of informal reports and seminars
Edited by A. Dold, Heidelberg and B. Eckmann, Zürich

171

Set-Valued Mappings, Selections and Topological Properties of 2^X

Organizing Committee
K. Kuratowski (Warsaw), G. S. Young (New Orleans),
S. B. Nadler, Jr. (Buffalo)

Proceedings of the conference held at
the State University of New York at Buffalo
May 8-10, 1969

Edited by W. M. Fleischman, Suny at Buffalo/USA

Springer-Verlag
Berlin · Heidelberg · New York 1970

ISBN 3-540-05293-3 Springer-Verlag Berlin · Heidelberg · New York
ISBN 0-387-05293-3 Springer-Verlag New York · Heidelberg · Berlin

Offsetdruck: Julius Beltz, Weinheim/Bergstr.

This volume is dedicated to Professor Kazimierz Kuratowski whose abundant contributions to general topology and the foundations of mathematics have greatly enriched the store of mathematical knowledge.

INTRODUCTION

This volume comprises the proceedings of the International Conference on Set-Valued Mappings, Selections, and Topological Properties of 2^X which met during the period from May 8 to May 10, 1969 at the State University of New York Campus at Buffalo under the sponsorship of the Department of Mathematics of that school. The director of the conference was Professor Kazimierz Kuratowski of the Polish Academy of Sciences. The Organizing Committee consisted of Professors Kuratowski, G. S. Young (New Orleans) and S. B. Nadler, Jr. (Buffalo).

Morning and afternoon sessions were held on each of the three days of the conference. At the first of these, chaired by Professor K. D. Magill, Jr. (Buffalo), the participants were welcomed by President Martin Meyerson of the State University of New York at Buffalo. Subsequent sessions were chaired by Professor A. H. Stone (Rochester) and S. B. Nadler, Jr., (Buffalo). Contributed papers generally averaged twenty minutes in length with an equal amount of time allotted between talks for questions and discussion. At the morning session of May 9, Professor E. A. Michael (Seattle) presented an invited forty-five minute survey of the theory of selections.

The significance of the conference is to be found equally in its summation of the status of many classical investigations in the areas of set-valued maps, semi-continuous decompositions, selections, and hyperspaces, and in its revelation of a number of new directions being taken by researchers particularly in the application of these ideas to analysis and algebraic topology. It is hoped that the present volume will be profitable to specialists in these areas as well to all mathematicians interested in the present state of the subject.

On the night of May 7, the participants in the conference were welcomed at a party given in their honor by the Department of Mathematics at the State University of New York at Buffalo. At the concluding luncheon on May 10,

Professor Kuratowski gave an informal talk on the history of the Polish school of mathematics.

The Organizing Committee wishes to express its thanks to all authors having forwarded their papers and also to all those at the SUNY at Buffalo who helped to bring about the conference. A particular measure of gratitude is due to Miss Anna Sheehan who patiently and accurately prepared the final typescript of these proceedings, to Mr. Robert Padian who proofread many of the manuscripts, and to Mrs. Irene V. Young and Miss Hilda Kurland, who as administrative assistants for the Department of Mathematics during the past two years rendered, expertly and graciously, indispensable assistance in the successful conduct of the conference and preparation of this volume.

Finally, we wish to express our appreciation to Springer-Verlag for undertaking the publication of these proceedings.

William Fleischman
State University of New York at Buffalo

List of Participants

A. R. Bednarek	University of Florida, Gainesville, Florida
Harold Bell	Case Western Reserve University, Cleveland, Ohio
Michael D. Bierbauer	Marymount College, Westchester, New York
Louis J. Billera	Cornell University, Ithaca, New York
Ken Bowen	Syracuse University, Syracuse, New York
Bruno Brosowski	Max-Planck Institut for Physics and Astrophysics, Munich
Leon Brown	Wayne State University, Detroit, Michigan
W. Comfort	Wesleyan University, Middletown, Connecticut
H. H. Covitz	State University of New York at Buffalo, Buffalo, New York
Ky Fan	University of California, Santa Barbara, California
O. Feichtinger	Montana State University, Bozeman, Montana
Henryk Fast	Wayne State University, Detroit, Michigan
Robert Fraser	Louisiana State University, Baton Rouge, Louisiana
George Goss	Wesleyan University, Middletown, Connecticut
N. Gray	Western Washington State College, Bellingham, Washington
B. Halpern	University of California, Berkeley, California
P. D. Hanke	University of Wisconsin, Milwaukee, Wisconsin
Robert Heath	University of Washington, Seattle, Washington
G. Henderson	University of Wisconsin, Milwaukee, Wisconsin
J. H. V. Hunt	University of Saskatchewan, Saskatoon, Saskatchewan, Canada
Gerald Itzkowitz	State University of New York at Buffalo, Buffalo, New York

Jan W. Jaworowski	University of Indiana, Bloomington, Indiana
James Keesling	University of Florida, Gainesville, Florida
Murray Kirch	State University of New York at Buffalo, Buffalo, New York
K. Kuratowski	State University of New York at Buffalo, Buffalo, New York
Norman Levine	Ohio State University, Columbus, Ohio
Vincent J. Mancuso	St. John's University, Jamaica, New York
B. L. McAllister	Montana State University, Bozeman, Montana
Louis F. McAuley	Rutgers University, New Brunswick, New Jersey
Ernest Michael	University of Washington, Seattle, Washington
S. Mrowka	State University of New York at Buffalo, Buffalo, New York
Sam B. Nadler, Jr.	State University of New York at Buffalo, Buffalo, New York
Togo Nishiura	Wayne State University, Detroit, Michigan
S. Patnaik	Universite de Montreal, Montreal, Quebec, Canada
M. J. Powers	Northern Illinois University, DeKalb, Illinois
C. J. Rhee	Wayne State University, Detroit, Michigan
P. Roy	State University of New York, Binghamton, New York
Michael M. Rubenstein	Case Reserve University, Cleveland, Ohio
E. Schneckenburger	State University of New York at Buffalo, Buffalo, New York
A. H. Stone	University of Rochester, Rochester, New York
K. Sundaresan	Carnegie-Mellon University, Pittsburgh, Pennsylvania
E. Tymchatyn	University of Saskatchewan, Saskatoon, Saskatchewan, Canada
Gerald Ungar	Case-Western Reserve University, Cleveland, Ohio
Giovanni Viglino	Wesleyan University, Middletown, Connecticut

John J. Walsh	Case Western Reserve University, Cleveland, Ohio
L. E. Ward	University of Oregon, Eugene, Oregon
Raymond Wong	University of California, Santa Barbara, California
Gail Young	Tulane University, New Orleans, Louisiana
Eufemio Yzazaga	Universidad Central de Nicaragua, Bluefields, Nicaragua

TABLE OF CONTENTS

FIXED POINTS, INDEX, AND DEGREE
FOR SOME SET VALUED FUNCTIONS

by Harold Bell

A topological space X is said to have the fixed point property if every continuous self mapping of the space leaves some point fixed, i.e. if $f: X \to X$ is continuous then $a = f(a)$ for at least one $a \in X$.

A continuous function $f: S^n \to S^n$ has degree k if for a "sufficiently close" simplicial approximation of f, say α, and some simplex S in the range space of α the set of simplexes in the domain space that map onto S is $\{A_1, A_2, \ldots, A_m\}$ and $k = \sum_{i=1}^{m} o(A_i)$, where $o(A_i)$ is the orientation of f restricted to A_i.

A continuous function $f: S^n \to R^{n+1}$ has index k with respect to the origin in R^{n+1}, 0 if $0 \notin f(S^n)$ and the degree of $g(z) = f(z)/_{\|z\|}$ is k. f has index k with respect to a point $p \in R^{n+1}$ if $g(z) = f(z) - p$ has index k with respect to 0.

It is the purpose of this paper to expand the above interrelated ideas with the use of semi-closure operators. The reader is referred to [4] for proofs of the theorems for the continuous cases which will be assumed here.

By a semi-closure operator on a topological space X we mean a set valued function T that assigns to every subset A of X a closed subset $T(A)$ of X for which 1) $A \subset T(A)$ for $A \subset X$, and

2) If $A \subset T(B)$ then $T(A) \subset T(B)$.

<u>Examples.</u>

1) $C(A) =$ the convex hull of A for X convex metric.

2) $T(A) = \overline{A} =$ the closure of A.

3) X is a continuum and $T(A)$ is the intersection of all subcontinua of X that contain A.

4) X is a continuum, $p \in X$, and $T(A) = X$ if $p \notin \bar{A}$ and $T(A)$ is the closure of the complement of the component of $X - \bar{A}$ that contains p.

Fixed point theory seems to have its origin in the intermediate value theorem of elementary calculus. We notice that the intermediate value theorem is dealt with easily with the purely topological notions of continuity and connectivity. However, the sufficiency of continuity and connectivity to deal with fixed points seems to end here. In fact all known proofs that the n-cell has the fixed point property for $n > 1$, resort to some form of discrete and indeed finite mathematics. The inductive step common to all proofs of the Brouwer fixed-point theorem, (as can be found in [9] for example) is "If a set with $n + 1$ elements is mapped into a set with n elements then either 0 or 2 proper subsets are mapped onto.".

A statement of the Brouwer theorem that would be more inclined to remove some of the mystery of the proof would be "Let X be the first $k + 1$ non-negative integers and let $f: X^n \to X^n$ for some positive integer n. Then there is a unit cube D such that any cube that contains $f(D \cap X^n)$ intersects D."

The apparent lack of cohesiveness between the idea of fixed points and the techniques used in the proofs becomes more apparent with the continuing production of counter examples to conjectures that attempt to use spaces that have the fixed point property to build new spaces with the fixed point property. Many of these counter examples are discussed in [2]. The purpose of this section of the paper is to suggest a way to broaden the approach to fixed point theory so as to incorporate most of the existing theory into a new theory that places the theory closer to the proofs and at the same time avoids some of the standard counter examples.

Definition: Let T be a semi-closure operator on a space Z and let $f: X \to Z$. Then $f_T: X \to 2^Z$ is defined by $f_T(x) = \cap\{T[f(U)]: x \in \text{int } U\}$. If A is a set we denote $\cup\{f_T(x): x \in A\}$ by $f_T(A)$. A space X has the fixed point property with respect to a semi-closure operator T if for $f: X \to X$ there is at least one $x \in X$ for which $x \in f_T(x)$. A function $r: X \to A$ is called a semi-retraction if $A \subset X$, $r(x) = x$ for $x \in A$, and r is continuous at each $a \in A$. A is called a semi-retract of X.

<u>Theorem</u>: (Eilenberg and Montgomery). Let f be an upper semi-continuous set valued function defined on a compact convex subset X of R^n. If $f(x)$ is an acyclic subcontinuum of X for each $x \in X$ then f has a fixed point.

<u>Theorem</u>: Suppose X has the fixed point property with respect to a semi-closure operator T and $A = T(A)$ is a semi-retract of X. Then A has the fixed point property with respect to T.

<u>Proof</u>: Let $f: A \to A$. Then $f(r): X \to A \subset X$ has a fixed point x_0. Since r is continuous at x_0 we have $f_T(x_0) \supset [f(r)]_T(x_0)$.

<u>Lemma</u>: Let T be a semi-closure operator on Y and let $f: X \to Y$. f_T is upper semi-continuous if each $x \in X$ has a neighborhood U for which $T(f(U))$ is compact.

<u>Proof</u>: Suppose $f_T(x) \subset V$ for some open set $V \subset Y$. Let U_0 be a neighborhood of x for which $T[f(U_0)]$ is compact. Then
$\emptyset = (f_T(x)) \cap (Y-V) = \cap \{(T[f(U \cap U_0)]) \cap (Y-V): x \in \text{int } U\}$. Since $(T[f(U \cap U_0)]) \cap (Y-V)$ is compact for $x \in \text{int } U$ there is a finite set of neighborhoods $\mathfrak{U}$ such that $\cap \{(T[f(U \cap U_0)]) \cap (Y-V): U \in \mathfrak{U}\} = \emptyset$.
If $U_1 = \cap \{U \cap U_0: U \in \mathfrak{U}\}$ then $(T[f(U_1)]) \cap (Y-V) = \emptyset$. We now have $T[f(U_1)] \subset V$. Therefore, $f_T(U_1) \subset V$.

<u>Theorem</u>: Let T be a semi-closure operator on a compact space $X \subset R^n$ such that $T(A)$ is acyclic for $A \subset X$. Then X has the fixed point property with respect to T.

<u>Proof</u>: Let $r(x): R^n \to X$ so that $\|r(x) - x\|$ is minimal for $x \in R^n$. r is clearly a semi-retraction.

Applications:

1) Let X be a treelike continuum and let $T(A)$ be the intersection of all subcontinua of X that contain A. Then X has the fixed point property with respect to T.

2) Let X be the 2-simplex in R^2 and let $T(A)$ be the intersection of all compact subsets of X that contain A and have connected complements in R^2.

The interest here is in the class of functions that preserve connected sets. That is, if $f(A)$ is connected when A is connected then there is an $x \in X$ such that every compact set that contains the image of a neighborhood of x and does not separate R^2 contains x. O. H. Hamilton in [6] proved that if X is the n-cell and $f: X \to X$ is such that the graph of connected sets is connected then f has a fixed point. However, the image of connected sets being connected does not guarantee a fixed point even in the case $n = 1$.

3) X is a compact, convex subset of R^n. $T(A)$ is the closed convex hull of A. This is easily equivalent to the Kakutani fixed point theorem.

Definition: Let T be a semi-closure operator on a space X and let $\mathcal{U}$ be a collection of subsets of X. Two functions $f, g: Y \to X$ are said to be homotopic with respect to T and $\mathcal{U}$ if there is a function $F; Y \times [0,1] \to X$ such that $F[(y,0)] = f(y)$ and $F[(y,1)] = g(y)$ for $y \in Y$ and $F_T[(y,s)]$ is contained in some $U_{(y,s)} \in \mathcal{U}$ for $(y,s) \in Y \times [0,1]$. If $\mathcal{U} = \{U\}$ we shall say f and g are homotopic with respect to U. We write $f \sim_T g$ where $\mathcal{U}$ or U is understood. We write $f \sim g$ to mean f and g are homotopic in the usual sense.

Lemma: $f \sim_T g$ is an equivalence relation on $\{h: h: Y \to X$ and $h_T(y) \in U_y \in \mathcal{U}\}$ for $y \in Y$.

Definition: For n a fixed positive integer and $p \in R^{n+1}$ define $H_p^n = \{f: f: S^n \to R^{n+1}$ and $p \notin f_C(S^n)\}$ where $C(A)$ is the closed convex hull of A. Here $\mathcal{U} = \{R^{n+1} - \{p\}\}$.

Lemma: If $f \in H_p^n$ then there is a $g \in G_p$ such that g is continuous and $f \sim_C g$.

Proof: Let $\mathcal{V}$ be an open cover of S^n for which $p \notin C[f(V)]$ for $V \in \mathcal{V}$. Let $\mathcal{J}$ be a simplical decomposition of S^n for which each n-cell in $\mathcal{J}$ is contained in some $V \in \mathcal{V}$. Let g be the map obtained by restricting f to the vertices of $\mathcal{J}$ and extending this restriction linearly on each n-cell of $\mathcal{J}$. $g \sim_C f$ by $F[(x,s)] = (g(x))s + (1-s)f(x)$.

Lemma: If $f,g \in H_p^n$ are continuous then $f \sim_C g$ iff $f \sim g$ in $R^{n+1} - \{p\}$.

Proof: Trivially if $f \sim g$ in $R^{n+1} - \{p\}$ then $f \sim_C g$. If $f \sim_C g$ with homotopy $F: S^n x[0,1] \to R^{n+1}$ then let $\mathcal{J}$ be a simplical decomposition of $S^n \times [0,1]$ for which $p \notin C[F(S)]$ for $S \in \mathcal{J}$. Then restrict F to $S^n \times \{0,1\} \cup V$ where V is the set of vertices of $\mathcal{J}$. Then re-extend F to $S^n \times [0,1]$ so that each $S \in \mathcal{J}$ is mapped into $C[F(S)]$.

Definition: If $f \sim_C g \in H_p^n$ and g is continuous we define the index of f to be the index of g with respect to p.

Lemma: $f \sim_C g$ iff f and g have the same index.

Definition: Let $K^n = \{f: f: S^n \to S^n$ and $0 \notin f_C(S^n)\}$. Let C' be the semi-closure operator on S^n defined by $C'(A) = \{z/_{\|z\|}: z \in C(A)\}$ if $0 \notin C(A)$ and $C'(A) = S^n$ if $0 \in C(A)$. For $f \in K$ the degree of f is defined to be the index of f with respect to C and $R^{n+1} - \{0\}$.

Lemma: If $\mathfrak{U}$ is the collection of proper subsets of S^n then $f \sim_{C'} g$ with respect to $\mathfrak{U}$ is an equivalence relation on K^n.

Lemma: $f \in K^n \to f \sim_{C'} g$ for some continuous function g.

Lemma: $f \sim_{C'} g$ iff f and g have the same degree.

Theorem: For $f \in K^n$ the following are equivalent:

1) The degree of f is zero.

2) $f \sim_{C'} g$ for some constant function g.

3) There is an extension F of f to then n + 1 unit ball, E^{n+1}, such that $0 \notin F_C(E^{n+1})$, i.e. $F_{C'}(x)$ is defined for $x \in E^{n+1}$.

<u>Conjecture</u>: Let $f: S^n \to S^n$ have degree $k \neq 0$. Let $D_f = \{x \in S^n:$ cardinal $f^{-1}_{C'}(x) < |k|\}$ $(f^{-1}_{C'}(x) = \{z: x \in f_{C'}(z)\} \emptyset)$. Then

1) The interior of D_f is empty.

2) The dimension of $D_f \leq n - 2$.

2) is of course stronger than 1).

In the case where f is continuous, light, interior and locally sense preserving 1) is known. (See [8]).

In the case where f is continuous and n = 2, D_f is finite (See [1]). In fact if D_x is defined to be the maximum of $\{K -$ cardinal $f^{-1}_{C'}(x)$ and $0\}$, $\sum_{x \in S^2} D_x \leq 2(|K| - 1)$. The proof used in the continuous case contains the proof for the case $f \in K^2$.

BIBLIOGRAPHY

1. H. Bell and P. Frederickson, On topological degree on the two sphere. To appear.

2. R. H. Bing, The elusive fixed point property, Amer. Math. Monthly, Vol. 76, Number 2, February 1969.

3. K. Borsuk, Sur un continu acyclique qui se laisse transformer topologiquent et lui même sans points invariants, Fund. Math. 24, (1935) pp. 51-58.

4. J. Cronin, Fixed points and topological degree in non-linear analysis. A.M.S. Mathematical Surveys, Number 11, 1964.

5. S. Eilenberg and D. Montgomery, Fixed point theorems for multivalued transformations, Amer. J. Math. 68 (1946) pp. 214-222.

6. O. H. Hamilton, Fixed points for certain noncontinuous transformations. A.M.S. Proceedings, Vol. 8, 1957, pp. 750-756.

7. S. Kakutani, A generalization of Brouwer's fixed-point theorem, Duke Math. J. 8 (1941) pp. 457-459.

8. C. J. Titus and G. S. Young, The extension of interiority with some applications. Trans. Amer. Math. Soc. 103 (1962) pp. 329-340.

9. G. T. Whyburn, Analytic topology. A.M.S. Colloquium Publications, Vol. XXVIII, 1940.

TOPOLOGICAL WELL-ORDERING AND CONTINUOUS SELECTIONS

by

R. Engelking, R. W. Heath and E. Michael[1)]

In this paper we introduce the concept of a topologically well-ordered subspace of a linearly ordered space, and prove that every complete, 0-dimensional metric space is homeomorphic to such a subspace. We apply this theorem to give a simple proof of a selection theorem which was recently announced by M. Čoban, and we also prove some related results.

By a linearly ordered space, we shall mean a space with a linear ordering, equipped with the usual order topology. A subspace S of such a space will always carry the induced subspace topology, not the (generally coarser) order topology obtained by considering S itself as a linearly ordered space.

DEFINITION. If X is a linearly ordered space, then a subspace A of X is called a topologically well-ordered subspace of X if every non-empty relatively closed subset of A has a first element.

We are now ready to state our main result, where dimension means the covering dimension dim (or, equivalently for metric spaces, the large inductive dimension Ind).

THEOREM 1.1. Every 0-dimensional, complete metric space X can be embedded as a closed, topologically well-ordered subspace of a linearly ordered space L.

It is evident from our proof that the space L in Theorem 1.1 can itself be chosen to be a complete, 0-dimensional metric space. In fact if X has weight m, then L can be chosen homeomorphic to the Baire space $B(m)$ (the

---

1) This is a summary of a paper appearing in Inventiones mathematicae Volume 6, pp. 150-158. The authors were all partially supported by an N.S.F. contract.

product of countably many discrete spaces of cardinality m).

We now apply Theorem 1.1 to prove a selection theorem which was recently announced by M. Čoban. If X is any topological space, then $\mathfrak{F}(X)$ denotes the collection of all non-empty closed subsets of X, equipped with the Vietoris topology. (That topology is generated by the subbase consisting of all collections of the form $\{A \in \mathfrak{F}(X): A \subset U\}$ or $\{A \in \mathfrak{F}(X): A \cap U \neq \emptyset\}$, with U open in X. For compact metric spaces X, the Vietoris topology agrees on $\mathfrak{F}(X)$ with the Hausdorff metric.). A <u>continuous</u> <u>selection</u> on a subspace $\mathfrak{A} \subset \mathfrak{F}(X)$ is a continuous map $f: \mathfrak{A} \to X$ such that $f(A) \in A$ for every $A \in \mathfrak{A}$. It is easily checked that, if X is a topologically well-ordered subspace of a linearly ordered space, then the map which assigns to each $A \in \mathfrak{F}(X)$ its first element is a continuous selection on $\mathfrak{F}(X)$. We therefore have the following corollary to Theorem 1.1.

COROLLARY 1.2. If X is a 0-dimensional, complete metric space, then there exists a continuous selection on $\mathfrak{F}(X)$.

Čoban asserts that his proof of Corollary 1.2 uses successive approximations; thus his proof is quite different from ours.

The statements of Theorem 1.1 and Corollary 1.2 appear to be the best possible in several respects. Let us briefly look at some possibilities.

The assumption that $\dim X = 0$ cannot be dropped, or even weakened to $\dim X \leq 1$. In fact, the circle cannot be embedded in any linearly ordered space, and Corollary 1.2 is also false in this case. (It is easy to check that there exists no continuous selection on the collection of all diametrically opposite two-point subsets of the circle). If X is the real line, then X can of course be linearly ordered, but we can show that Corollary 1.2, and <u>a fortiori</u> Theorem 1.1, are false in this case as well.

The completeness of X is indispensable in Theorem 1.1. In fact we have shown that any metric space which can be embedded as a (not necessarily closed) topologically well-ordered subspace of a (possibly non-metrizable) linearly ordered space must be completely metrizable. In the case of Corollary 1.2, we have shown that the completeness hypothesis cannot be eliminated; indeed, we proved that, if Q denotes the rationals, there exists no continuous selection on $\mathfrak{F}(Q)$. We do not know, however, whether a metrizable space X which admits a continuous selection on $\mathfrak{F}(X)$ must always be completely metrizable.

Finally, one may ask whether the conclusion of Theorem 1.1 can be strengthened -- and simplified -- by taking $L = X$. This may appear plausible, since it was proved by H. Herrlich that every 0-dimensional metric space is homeomorphic to a linearly ordered space L. Nevertheless, it cannot always be done. In fact we have shown that one can take $L = X$ in Theorem 1.1 if and only if X is both locally compact and separable.

References

1. N. Bourbaki, Topologie Générale, Chapter 4, Hermann, Paris, 1960.

2. R. Engelking, Outline of General Topology, North-Holland, Amsterdam, 1968.

3. H. Herrlich, Ordnungsfähigkeit total-diskontinuierlicher Räume, Math. Ann. 159 (1965), 77-80.

4. E. Michael, Topologies on spaces of subsets, Trans. Amer. Soc. 71 (1951), 152-182.

EXTENSIONS OF TWO FIXED POINT THEOREMS OF F. E. BROWDER

by Ky Fan

1. Single-valued mappings. For single-valued mappings, we have the following new fixed point theorem.

(1) Let X be a non-empty compact convex set in a locally convex, Hausdorff topological vector space E, and let $f: X \to E$ be an arbitrary continuous mapping from X into E. Then either f has a fixed point in X, or there exist a point $y_0 \in X$ and a continuous semi-norm p on E such that

$$\min_{x \in X} p(x - f(y_0)) = p(y_0 - f(y_0)) > 0.$$

From (1) one can easily derive

(2) Let X be a non-empty compact convex set in a locally convex Hausdorff topological vector space E, and let $f: X \to E$ be a continuous mapping. If for each $y_0 \in X$, there is a number λ (real or complex depending on whether the vector space E is real or complex) such that $|\lambda| < 1$ and $\lambda y_0 + (1-\lambda) f(y_0) \in X$, then f has a fixed point in X.

For a convex set X in a real vector space E, the algebraic boundary $\delta(X)$ of X is defined as the set of those $x \in X$ for which there exists a $y \in E$ such that $x + \lambda y \notin X$ for all $\lambda > 0$. (2) is closely related to the following theorem.

(3) (Browder [1]). Let X be a non-empty compact convex set in a locally convex, Hausdorff topological vector space E, and let $f: X \to E$ be a continuous mapping. If for every point y_0 in the algebraic boundary $\delta(X)$ of X, there exist a point $x_0 \in X$ and a real number $\mu > 0$ such that $f(y_0) - y_0 = \mu(x_0 - y_0)$, then f has a fixed point in X.

As was pointed out by Browder [1], the last theorem remains valid if "$\mu > 0$" is replaced by "$\mu < 0$". It is clear that each of (1), (2), (3) generalizes the classical fixed point theorem of Tychonoff [11].

2. _Upper demi-continuous set-valued mappings._ Let E be a real Hausdorff topological vector space, and let $X \subset E$. Let f be a set-valued mapping defined on X such that for each $x \in X$, $f(x)$ is a non-empty subset of E. As usual, f is said to be _upper semi-continuous_ on X, if for every point $x_0 \in X$ and any open set G in E containing $f(x_0)$, there is a neighborhood N of x_0 in X such that $f(x) \subset G$ for all $x \in N$. To introduce the more general concept of upper demi-continuity, we recall that an _open half-space_ H in E is a set of the form $H = \{x \in E: \varphi(x) > r\}$, where φ is a continuous linear form on E, not identically zero, and r is a real number. We shall say that f is _upper demi-continuous_ on X, if for every $x_0 \in X$ and any open half-space H in E containing $f(x_0)$, there is a neighborhood N of x_0 in X such that $f(x) \subset H$ for all $x \in N$. As usual, we say that two sets A, B in E can be _strictly separated by a closed hyperplane_, if there exist a continuous linear form φ on E and a real number r such that $\varphi(x) < r$ for $x \in A$, and $\varphi(y) > r$ for $y \in B$.

(4) _Let X be a non-empty compact convex set in a real, Hausdorff topological vector space E. Let f, g be two upper demi-continuous mappings defined on X such that for each $x \in X$, $f(x)$ and $g(x)$ are non-empty subsets of E. If for every point $x \in \delta(X)$ there are three points $y \in X$, $u \in f(x)$, $v \in g(x)$ and a real number $\lambda > 0$ such that $y - x = \lambda(u - v)$, then there exists a point $\hat{x} \in X$ for which $f(\hat{x})$ and $g(\hat{x})$ cannot be strictly separated by a closed hyperplane._

By exchanging the roles of f and g, one sees that (4) remains valid if "$\lambda > 0$" is replaced by "$\lambda < 0$". As a corollary of (4), we have

(5) *Let* X *be a non-empty compact convex set in a real, Hausdorff topological vector space* E. *Let* f, g *be two upper demi-continuous mappings defined on* X *such that for each* $x \in X$, $f(x)$ *and* $g(x)$ *are non-empty subsets of* E. *Suppose that the following two conditions are satisfied:*

(5.1) *For each* $x \in \delta(X)$, *there exist two points* $y \in X$, $u \in f(x)$ *and a real number* $\lambda > 0$ *such that* $y - x = \lambda(u - x)$.

(5.2) *For each* $x \in \delta(X)$, *there exist two points* $z \in X$, $v \in g(x)$ *and a real number* $\mu < 0$ *such that* $z - x = \mu(v - x)$.

Then there exists a point $\hat{x} \in X$ *for which* $f(\hat{x})$ *and* $g(\hat{x})$ *cannot be strictly separated by a closed hyperplane.*

In (4) and (5), the topological vector space E is not required to be locally convex, also the sets $f(x)$, $g(x)$ need not be convex, nor closed. In case E is locally convex, one can combine (4) or (5) with the classical theorem on separation of convex sets to obtain results on coincidences of f, g. Thus (4) implies the following result.

(6) *Let* X *be a non-empty compact convex set in a real, locally convex Hausdorff topological vector space* E. *Let* f, g *be two upper demi-continuous set-valued mappings defined on* X *such that for each* $x \in X$, $f(x)$, $g(x)$ *are non-empty closed convex sets in* E *and at least one of them is compact. If for every* $x \in \delta(X)$, *there are three points* $y \in X$, $u \in f(x)$, $v \in g(x)$ *and a real number* $\lambda > 0$ *such that* $y - x = \lambda(u - v)$, *then there exists a point* $\hat{x} \in X$ *for which* $f(\hat{x})$ *and* $g(\hat{x})$ *have a non-empty intersection.*

If g is the identity mapping of X, (6) specializes to the following theorem of Browder [2], who used upper semi-continuity instead of upper demi-continuity.

(7) (Browder [2]). *Let* X *be a non-empty compact convex set in a real, locally convex, Hausdorff topological vector space* E. *Let* f *be an upper demi-continuous set-valued mapping defined on* X *such that for each* $x \in X$, $f(x)$ *is a non-emtpy closed convex set in* E. *Suppose that for each* $x \in \delta(X)$ *we can find two points* $y \in X$, $u \in f(x)$ *and a real number* $\lambda > 0$ *such that* $y - x = \lambda(u - x)$. *Then there exists a point* $\hat{x} \in X$ *with* $\hat{x} \in f(\hat{x})$.

This theorem (7) unifies (3) and an earlier generalization [3], [7] of Kakutani's fixed point theorem [9].

3. *A lemma.* In the proofs [6] of our main results (1) and (4), we make use of the following lemma.

(8) *Let* X *be a non-empty compact convex set in a Hausdorff topological vector space. Let* A *be a closed subset of* $X \times X$ *having the following properties:*

(8.1) $(x, x) \in A$ *for all* $x \in X$.

(8.2) *For each* $y \in X$, *the set* $\{x \in X: (x,y) \notin A\}$ *is convex (or empty).*

Then there exists a point $y_0 \in X$ *such that* $X \times \{y_0\} \subset A$.

This lemma was given in our paper [4] and its proof was based on a well-known classical theorem of Knaster-Kuratowski-Mazurkiewicz [10]. Other applications of (8) have been made by the writer [5], Iohvidov [8], and Wittstock [12].

References

[1] F. E. Browder, A new generalization of the Schauder fixed point theorem, Math. Annalen 174 (1967), 285-290.

[2] F. E. Browder, The fixed point theory of multi-valued mappings in topological vector spaces, Math. Annalen 177 (1968), 283-301.

[3] K. Fan, Fixed-point and minimax theorems in locally convex topological linear spaces, Proc. Nat. Acad. Sci. U.S. 38 (1952), 121-126.

[4] K. Fan, A generalization of Tychonoff's fixed point theorem, Math. Annalen 142 (1961), 305-310.

[5] K. Fan, Invariant subspaces of certain linear operators, Bull. Amer. Math. Soc. 69 (1963), 773-777.

[6] K. Fan, Extensions of two fixed point theorems of F. E. Browder, Math. Z. 112 (1969), 234-240.

[7] I. L. Glicksberg, A further generalization of the Kakutani fixed point theorem, with application to Nash equilibrium points, Proc. Amer. Math. Soc. 3 (1952), 170-174.

[8] I. S. Iohvidov, On a lemma of Ky Fan generalizing the fixed-point principle of A. N. Tihonov (in Russian), Dokl. Akad. Nauk SSSR 159 (1964), 501-504; English translation: Soviet Math. 5 (1964), 1523-1526.

[9] S. Kakutani, A generalization of Brouwer's fixed-point theorem, Duke Math. J. 8 (1941), 457-459.

[10] B. Knaster, C. Kuratowski and S. Mazurkiewicz, Ein Beweis des Fixpunkt-satzes für n-dimensionale Simplexe, Fund. Math. 14 (1929), 132-137.

[11] A. Tychonoff, Ein Fixpunktsatz, Math. Annalen 111 (1935), 767-776.

[12] G. Wittstock, Über invariante Teilräume zu positiven Transformationen in Räumen mit indefiniter Metrik, Math. Annalen 172 (1967), 167-175.

PROPERTIES OF THE λ TOPOLOGY

(Preliminary Report)

by

Oskar Feichtinger

1. Introduction. Certain topologies on the space of all nonvoid closed subsets of a topological space X have been examined in considerable detail. For example, E. Michael and K. Kuratowski investigated the finite (or exponential topology [4, 2], whereas V. I. Ponomarev [5] studied the so-called κ topology. In most cases restrictions have been placed on either the space X or on the collection of subsets itself. We have tried to extend some of the published results and to generalize and expand on topics pertaining to the "λ-topology" on spaces of nonvoid subsets (not always closed) of X.

2. Definitions. If $\mathcal{C}$ is a family of subsets of X, a binary relation R can be defined on X to $\mathcal{C}$ as follows: $(x, C) \in R$ iff $x \in C$, where $C \in \mathcal{C}$. For $A \subseteq X$, let R_+ and $R_-: \mathcal{P}(X) \to \mathcal{P}(\mathcal{C})$ be defined by $R_+(A) = \{C \in \mathcal{C} | C \subseteq A\}$, $R_-(A) = \{C \in \mathcal{C} | C \text{ meets } A\}$. R^+ and R^- will denote mappings of $\mathcal{P}(\mathcal{C})$ to $\mathcal{P}(X)$, where for $\mathcal{A} \subseteq \mathcal{C}$, $R^-(\mathcal{A}) = \bigcup\mathcal{A}$ and $R^+(\mathcal{A}) = X - R^-(\mathcal{C} - \mathcal{A}) = \{x \in X |$ if $x \in C \in G$ then $C \in \mathcal{A}\}$. Using these multifunctions, we let $\lambda\mathcal{C}$ be the space $\mathcal{C}$ with the following topology: the open sets are generated by sets of the form $R_-(G)$ for G open in X. With this convention, no distinction will be made between $\mathcal{C}$ and $\lambda\mathcal{C}$ unless ambiguities might occur. R is said to be closed if for any K closed in X, $R_-(K)$ is closed in $\lambda\mathcal{C}$. It is easy to show that R closed is equivalent to R_+ (open) is open. For convenience's sake, we shall denote by λK the space of all nonvoid closed subsets of X and by $\lambda\mathcal{P}$ the space of all nonvoid subsets of X, with the λ topology.

3. Let $\mathcal{C}$ be a family of subsets of X including all sets $\{x\}$, for $x \in X$.

Proposition 1. X is compact iff λC compact.

Proof: Let $\{R_-(G_\alpha)\}$ be a cover of λC by subbasic open sets. Then $\{G_\alpha\}_\alpha$ covers X, since for every $x \in X$, $\{x\} \in R_-(G_\alpha)$ (some α). By X compact, $\{G_i\}_{i=1}^n$ covers X and $\{R_-(G_i)\}_{i=1}^n$ covers λC, since every element of C must meet at least one G_i. The converse is proven in a similar manner.

If $X \in \lambda C$, then $X \in R_-(G)$ for every $G \subseteq X$, G open; hence if $X \in \lambda C$, then λC is separable. (Of course the implication λC separable $\rightarrow X$ separable is not true).

Proposition 2. X is second countable $\leftrightarrow \lambda C$ is second countable.

Proof: $\rightarrow$: Let $\mathcal{O} = \{O_1, O_2, \ldots\}$ be a countable base for X. Let $\mathcal{B}$ be the family of all finite intersections of the form $R_-(O_i)$ with $O_i \in \mathcal{O}$. Then $\mathcal{B}$ is a countable base for λC.

$\leftarrow$: Since λC has a countable base, we can extract a countable base of the form $R_-(E_1^j) \cap \ldots \cap R_-(E_{n_j}^j)$, $j = 1, 2, \ldots$ Let $O \subseteq X$ be open, $x \in O$. Then $\{x\} \in R_-(O)$ and $\{x\} \in R_-(E_1^j) \cap \ldots \cap R_-(E_n^j) \subseteq R_-(O)$ for some j. Thus $x \in E_1^j \cap \ldots \cap E_n^j \subseteq O$. Hence the E_i^j together with all their finite intersections form a countable base for X.

Proposition 3. Let X be first countable, C a family of countable subsets of X. Then λC is first countable.

Proof: Let $A \in C$, $A = \{x_1, x_2, \ldots\}$. For all x_i, let $\{O_{ij}\}$ $(j = 1,2,..)$ be a countable point base. Form $\{R_-(O_{ij}) \mid i = 1, 2, \ldots ; j = 1, 2, \ldots\}$. Let $A \in R_-(G)$, G open. Then some $x_k \in A \cap G$, hence there is an $O_{ik} \subseteq G$ such that $x_k \in O_{ik}$. Then $R_-(O_{ik}) \subseteq R_-(G)$. Thus finite intersections of the $R_-(O_{ij})$ form the required local base.

A related theorem can be proven in a similar manner:

Corollary: Let X be a nonseparable metric space, $\mathcal{C}$ a family of Lindelöf subsets. Then $\lambda\mathcal{C}$ is first countable.

Proposition 4. If $\lambda\mathcal{C}$ is first countable, then so is X. ($\mathcal{C}$ a family of subsets containing all singletons).

Proof: Let $\{x\} \in \lambda\mathcal{C}$. There is a countable neighborhood base of the form $\{R_-(N_1^i) \cap \ldots \cap R_-(N_{ki}^i)\}$. Let N be any open neighborhood of x in X. Then $R_-(N)$ is an open neighborhood of $\{x\}$ and $\{x\} \in R_-(N_1^i) \cap \ldots \cap R_-(N_{ki}^i) \subseteq R_-(N)$ for some i. Hence $x \in N_1^i \cap \ldots \cap N_{ki}^i \subseteq N$, and the N_i^j together with finite intersections form a countable neighborhood base.

Proposition 5. If X is T_1, then elements $\{x\}$, $x \in X$, are the only points of λX without limit points in λX.

Proof: Let $A \in \lambda X$. Then $A \in R_-(X - \{x\})$ and $\{x\} \notin R_-(X - \{x\})$, $(\{x\} \neq A)$ hence A is not a limit point of $\{\{x\}\}$. If $x \in B \in \lambda X$ and B is not a singleton, then $\{x\}$ is a limit point of $\{B\}$. (This is also true for any $\lambda\mathcal{C}$ where $\mathcal{C}$ includes all the singleton subsets of X).

The above theorem is the tool required to prove the following

Proposition 6. Let X, Y be T_1 spaces. Then X is homeomorphic to Y iff λX is homeomorphic to λY.

Proof: If X and Y are homeomorphic, the result follows trivially. The following statement reformulates the converse.

Lemma: Knowledge of λX and its topology completely determines X and the topology on X.

Proof: X is the union of the set of closed points of λX. Let $\mathcal{A}$ be open in λX. Form the union of the set of singletons belonging to $\mathcal{A}$. The resulting set is open in X. Furthermore, even open set G in X can be obtained by taking

$\mathcal{Q} = R_{-}(G)$.

Several results using properties of connectedness may be worth mentioning.

Proposition 7: Let $\mathcal{B} \subseteq \lambda C$ be such that $\mathcal{B}$ is connected in λC and let each element of $\mathcal{B}$ be connected in X. Then $R^{-}(\mathcal{B})$ is connected.

Proof: Assume that $R^{-}(\mathcal{B})$ is not connected, i.e., $R^{-}(\mathcal{B}) = H \cup K$(sep.), where $H = H' \cap R^{-}(\mathcal{B})$, $K = K' \cap R^{-}(\mathcal{B})$ with K', H' open in X. Then $\mathcal{H} = R_{-}(H') \cap \mathcal{B}$ and $\mathcal{K} = R_{-}(K') \cap \mathcal{B}$ are clearly open subsets of $\mathcal{B}$ and $\mathcal{H} \cup \mathcal{K} = \mathcal{B}$ (sep.), a contradiction.

The following example shows that if not all of the elements of $\mathcal{B}$ are connected, the theorem does not hold. Let $X = \{a, b\}$, let the topology τ on X be $\{\emptyset, \{a\}, \{b\}, \{a,b\}\}$, and let $C = \tau - \{\emptyset\}$. The open sets of λC are $\{\{a\}, \{a,b\}\}$; $\{\{b\}, \{a,b\}\}$; τ; $\{\{a,b\}\}$.
Clearly C is connected, but $R^{-}(C) = \cup C = X$ is not.

Corollary: If $\mathcal{Q} \subseteq C$ is a cover of X by connected sets and if $\mathcal{Q}$ is connected in λC, then X is connected.

Corollary: If X is not connected and if C is a covering of X by connected subsets of X then λC is not connected.

Proposition 8: Let C be any collection of subsets of X with $X \in C$. Then λC is connected.

Proof: Clear.

Consider now a multifunction on X to Y with the property that $\forall x \in X$, f(x) is closed in Y. Analogous to the definitions of R_{-}, R^{+}, etc., we define for $A \subseteq X$

$$f_-(A) = \{y \mid \exists x \in A \ni y \in f(x)\}$$

$$f_+(A) = Y - f_-(X - A)$$

and for $B \subseteq Y$,

$$f^-(B) = \{x \mid f(x) \cap B \neq \emptyset\}, \quad f^+(B) = \{x \mid f(x) \subseteq B\}.$$

The notation is largely due to Berge. See also [3]. With this convention, f is lower semicontinuous iff f^- takes open sets to open sets, or equivalently f^+ takes closed sets to closed sets. f is said to be closed if for $A \subseteq X$ closed, $f_-(A)$ is closed in Y. f gives rise to a single valued function $F: X \to \lambda Y$ by $F(x) = f(x)$. Our notation yields a particularly clean proof of

Proposition 9: f is lsc iff F is continuous.

Proof: Let $R_+(K)$ be a subbasic closed set in λY. Then $F^{-1}(R_+(K)) = \{x \mid f(x) \subseteq K\} = f^+(K)$.

If f is a closed multifunction, a further extension to $\overline{F}: \lambda X \to \lambda Y$ takes place, where for $A \in \lambda X$, $\overline{F}(A) = f_-(A)$. We give as intermediate step the following

Lemma: Let $B \subseteq Y$. $\overline{F}^{-1}(R_-(B)) = R_-(f^-(B))$ and $\overline{F}^{-1}(R_+(B)) = R_+(f^+(B))$.

Proof (part 1): $\overline{F}^{-1}(R_-(B)) = \{A \in \lambda X \mid \overline{F}(A) \cap B \neq \emptyset\} =$ $A \in \lambda X \mid f_-(A) \cap B \neq \emptyset\} = \{A \in \lambda X \mid A \cap f^-(B) \neq \emptyset\} = R_-(f_-(B))$.

We now have easily

Proposition 10: f is lsc $\leftrightarrow$ $\overline{F}$ is continuous.

Using Proposition 7 the proof of the next proposition can be supplied by the reader.

Proposition 11: Let $f: X \to Y$ be onto, lsc with closed connected point values. If X, Y are T_1 then if X is connected, Y is also connected.

Consider now the multifunction $f: X \to Y$ as a subset of $X \times Y$. Let $g: X \to X \times Y$ be defined by $g(x) = \{(x,y) | y \in f(x)\}$. Then g is the inverse of the restriction to f of the projection. Clearly g is open. It is not difficult to show that the next lemma holds.

Lemma: With g defined as above, g is lsc $\leftrightarrow$ f is lsc.

When f is single valued and continuous, the existence of a homeomorphism of $f \subseteq X \times Y$ to X is well known. (See Hall and Spencer [1]). Since $g(x)$ is a subset of $f \subseteq X \times Y$, it is reasonable to inquire under what conditions $\{g(x) | x \in X\}$ considered as a family of closed subsets of $\lambda(X \times Y)$ is homeomorphic to X. The following diagram shows the situation for f lsc, closed, on X onto Y, both T_1 spaces.

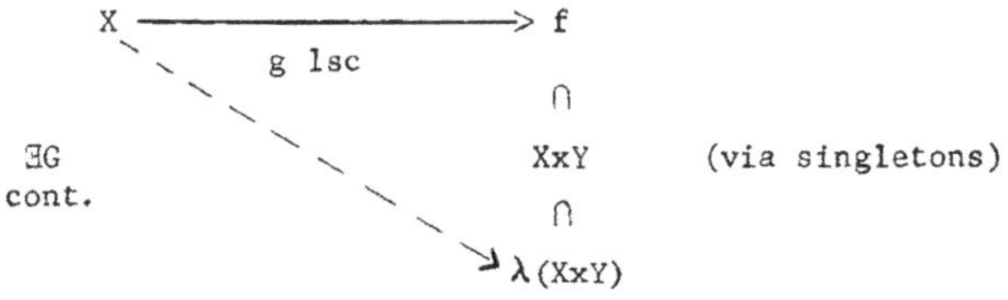

The existence and continuity of G follows from the preceding lemma and Proposition 9. G is clearly one-to-one onto $G(X)$. It remains to show that $G: X \to G(X)$ is closed (or open).

Let $U \subseteq X$ be closed. Then

$$G(U) = \{g(x) | x \in U\}, \quad \text{and} \quad R_+(U \times f_-(U)) \cap G(X) = \{g(x) | g(x) \subseteq U \times f_-(U)\} = G(U).$$

4. The above results are typical of our work with spaces of the form λC. In addition, work has been done on what the author called $\bar{\lambda}$ spaces, which are spaces with the same underlyings sets as those treated here, but in which basic open sets have the form $\bigcap_\alpha R_-(0_\alpha)$, where $\{0_\alpha\}$ forms a locally finite family of open sets. This modification was suggested by Professor R. Gillette.

REFERENCES

[1] Hall, D. W. and Spencer, G. L. II, Elementary Topology, Wiley and Sons (1955).

[2] Kuratowski, K., Topology, Vol. I., Academic Press (1966), 160-187.

[3] McAllister, B., The Application of a Function to Unions and Intersections of Sets, Mathematics Magazine, Vol. 42, No. 2, March 1969.

[4] Michael, E., Topologies on Spaces of Subsets, Transactions of the American Mathematical Society, Vol. 71, No. 1, 152-182, July 1951.

[5] Ponomarev, V. I., A New Space of Closed Sets and Multivalued Continuous Mappings of Bicompacta, American Mathematical Society Translations (ser. 2), Vol. 38, 95-118.

SEQUENCES OF CONTRACTIVE SET-VALUED MAPS

R. B. Fraser

In [3], S. Nadler proved several theorems on the behavior of fixed points for contraction maps. These results were extended in [4] to set-valued contraction maps.

Nadler and I were interested in what types of theorems could be proved if the mappings were only contractive maps instead of contraction maps. We proved the following

Theorem 1. Let (x, ρ_o) be a locally compact metric space. For each $n = 0, 1, \ldots,$ let ρ_n be a metric on X topologically equivalent to ρ_o and $f_n:(X, \rho_n) \to (X, \rho_n)$ a $(p_n\text{-})$ contractive mapping with fixed point a_n. If ρ_n converges uniformly to ρ_o and f_n converges pointwise to f_o then a_n converges to a_o.

We now indicate how this theorem can be extended to set-valued mappings. First we establish terminology.

Let X be a metric space. By 2^X, we mean the non-empty compact subsets of X. We metrize 2^X with the Hausdorff metric, which we denote by H. If F maps X into 2^X, then we extend F to a mapping $F: 2^X \to 2^X$ by $F(A) = \bigcup\{F(a) | a \in A\}$ for each $A \in 2^X$. Nadler showed in [4] that if F is a contractive (contraction) map, then F is a contractive (contraction) map.

A point x is a fixed point for a set-valued map F if $x \in F(x)$. The following result is useful in that it guarantees the existence of fixed points under reasonable circumstances.

Theorem 2. Let (x, ρ) be a metric space and $F: X \to 2^X$ a contractive map. If there exists $A, B \in 2^X$ such that some subsequence of $F^n(A)$ converges to B, then there exists $b \in B$ such that $b \in F(b)$; i.e. b is a fixed point for F.

We now present the modifications needed to extend Theorem 1 to the set-valued case. Let (X, ρ_o) be locally compact space. For each $n = 0, 1, 2, \ldots$, let ρ_n be a metric on X and H_n the Hausdorff metric on 2^X. We assume each ρ_n is topologically equivalent to ρ_o . Then a result of E. Michael's [2; Theorem 3.3] implies that each H_n is topologically equivalent to H_o. Let ρ_n converge uniformly to ρ_o . Standard computations show that H_n converges uniformly to H_o .

For each non-negative integer n, suppose that A_n is an element of 2^X such that for the contractive mapping $F_n: (X, \rho_n) \to (2^X, H_n)$, $F_n(A_n) = A_n$. (Such a set exists if the conditions of Theorem 2 are satisfies.) If F_n converges pointwise to F_o, then F_n converges uniformly (with respect to H_o) to F_o on compact subsets of X.

Essentially, this is because the uniform convergence of the metrics, the fact that each F_n satisfies a Lipschitz condition, and the uniform equivalence of the ρ_n's on compact subsets collectively guarantee that the F_n's form an equicontinious family (on the given compact set). Thus F_n converges pointwise to F_o .

We now apply Theorem 1 to observe that A_n converges to A_o. A theorem of Michael [2; Theorem 2.5] shows that $\bigcup_{n=0}^{\infty} A_n$ is compact. Letting a_n be a fixed point for F_n $(n > o)$, there is a convergent subsequence a_{n_r}. Now $(\rho_o(a_o,F_o(x)) = \lim \rho_o(a_{n_r},F_{n_r}(x_{n_r})) = 0$ whence the limit point a_o is a fixed point for F_o. The "diagonal limit" is permissible since F_n converges uniformly to F_o on compact subsets.

These and other related results, as well as detailed proofs, will appear in a joint paper by Nadler and me in the Pacific Journal of Mathematics.

BIBLIOGRAPHY

1. M. Edelstein, "On fixed and periodic points under contractive mappings," Journal London Math. Soc., 37 (1962), pp. 74-79.

2. E. Michael, "Topologies on spaces of subsets," Trans. Amer. Math. Soc., 71 (1951), pp. 152-182.

3. S. B. Nadler, Jr., "Sequences of contractions and fixed points," Pacific J. of Math., 27 (1968), pp. 579-585.

4. ______________, "Multi-valued contraction mappings," Pacific J. of Math., 30 (1969), pp. 475-488

ALGEBRAIC TOPOLOGY AND SET-VALUED MAPS

by Benjamin Halpern

Introduction: We will be concerned with continuous set-valued maps $F:X \to Y$ from one topological space into another. We will consider in particular maps F such that for each $x \in X$ the number of points in $F(x)$ is an integer which lies in some preassigned set of integers S. For example, we will study maps F such that for each $x \in X$, $F(x)$ consists of two or three distinct points. Given a finite set of positive integers S, a continuous set-valued map $F:X \to Y$ such that for each $x \in X$, $\#F(x)$ = cardinality of $F(x) \in S$ is equivalent to a continuous single valued map $\overline{F}$ from X into $Y^S = \{A \in 2^Y | \#A \in S\}$ which is given the relative topology from 2^Y. This focuses our study of set-valued maps onto the study of spaces of the form Y^S.

Section 1: Polyhedral Properties of X^S.

We will let $X^{\underline{n}}$ denote $X^{\{1, 2, \ldots, n\}}$. By making use of the lexicographic order on $(I^m)^n$ where $I = [0,1]$, one is able to cut $(I^m)^{\underline{n}}$ up into nice cells in a very regular way. This method of subdivision extends to $X^{\underline{n}}$ for polyhedral X and ultimately results in showing that $X^{\underline{n}}$ is a finite polyhedron if X is. Furthermore, for an arbitrary finite subset of positive integers S the space $X^S \subset X^{\underline{n}}$ where $n = \max S$ and there exists a subpolyhedron P of a subdivision of $X^{\underline{n}}$ such that P is a strong deformation retract of X^S. One result of these methods of subdivision is the following calculation:

Section 2: The Homology of $G^{\underline{n}}$ for Any Finite Graph G.

Set $q = \dim H_1(G, Z)$, Z = the integers. For $p > 0$

$$H_p(G^{\underline{n}}; Z) = Z^{\alpha^n_{pq}} \quad \text{where}$$

$$\alpha^n_{pq} = \begin{cases} \beta_{pq} & \text{if} \quad p = n \quad \text{or} \quad n - 1 \\ 0 & \text{otherwise} \end{cases}$$

$$\beta_{pq} = \begin{cases} 1 + \sum\limits_{\substack{k\ \mathrm{odd} \\ k=1}}^{p} \binom{k+q-1}{k} & p \ \text{odd} \\ \sum\limits_{\substack{k\ \mathrm{even} \\ k=0}}^{p} \binom{k+q-1}{k} & p \ \text{even} \end{cases}$$

This extends the results of Wu [5].

We also can calculate i_* where i is the inclusion $i:G^{\underline{n}} \to G^{\underline{n+1}}$.
The only non-trivial case is for dimension n and we have

$$\begin{array}{ccc} H_n(G^{\underline{n}}) & \xrightarrow{i_*} & H_n(G^{\underline{n+1}}) \\ \simeq & & \simeq \\ Z^{\beta_{nq}} & \longrightarrow & Z^{\beta_{nq}} \\ \\ \Psi & & \Psi \\ x & \longrightarrow & 2x \end{array}$$

We also can show that $\pi_1(G^{\underline{n}}) = 0$ for $n \geq 3$ and from this and the above calculations on homology we can calculate some other homotopy groups using the Hurewicz theorem.

Possibly the most important application of our method of subdivision is:

Section 3: The Piecewise Linear Approximation Theorem.

Let X and Y be finite polyhedra and $F:X \to Y^s$ a continuous map. Then we can distort F through an arbitrarily small homotopy $H:X \times I \to Y^s$ to a nice map $F':X \to Y^s$ such that for a sufficiently fine subdivision of X we have for each simplex σ of X, n single valued linear maps $g_i : \sigma \to Y$ $i = 1, \ldots, n$, $n = \max S$, such that $F'(x) = \{g_1(x), g_2(x), \ldots, g_n(x)\}$ for all $x \in \sigma$.

The significance of this theorem is that we can study arbitrary continuous set-valued maps $F : X \to Y^S$ by nice maps F'. These latter maps can be handled in a combinatorial manner. Indeed, the piecewise linear approximation theorem can be used effectively in making the following calculations.

Section 4: The Fundamental Group of M^S for All Manifolds M (dim $M \geq 2$) and some S.

	S	$\pi_1(M^S)$
	$\{n\}$	braid group if $M = I^2$
$n \geq 2$	$\{1,n\}$	$H_1(M; Z)$
$n > 2$, n even	$\{2,n\}$	$F(\frac{n}{2} - 1)$
n odd composite	$\{2,n\}$	$F(\frac{n-3}{2})$ f $Z/2$
n odd prime	$\{2,n\}$	$F(\frac{n-3}{2})$ f $(Z/2)$ A $H_1(M;Z/n)$
$m > n \geq 3$	$\{m,n\}$	$F(P(m,n)-1)$

where $F(q)$ = free group on q generators, f ≡ free product, A ≡ semi-direct product, $P(m,n)$ = the number of partitions of m into n positive parts.

An important homotopy invariant used in the above calculations is:

Section 5: The Induced Homology Homomorphism.

Consider a continuous map $F : X \to Y^{\{2,p\}}$ where X and Y are polyhedra and p is an odd prime. Under certain circumstances we can define an induced homology homomorphism $F_* : H(X; Z/p) \to H(Y, Z/p)$ which depends only on the homotopy class of F. The definition (which we shall not give) is based upon a method of assigning elements of Z/p to the points on the graph $\{(x,y) \in X \times Y \mid y \in F(x)\}$ of F in a "conservative" way. Such an assignment isn't always possible but there exists a homomorphism $C_F: H_1(X; Z) \to Z/p-1$ which measures the obstruction to such an assignment. In fact, F_* can be defined when $C_F = 0$. As a corollary, F_* can be defined if $H_1(X; Z) = 0$ or if X is simply connected. Besides aiding in the calculation of some fundamental groups, the induced homomorphism F_* (when definable) satisfies the Lefschetz theorem: If $\Lambda(F_*) \neq 0$ then there exists $x \in X$ such that $x \in F(x)$.

O'Neill, [4] defined an induced homomorphism G_* for continuous maps $G: X \to Y^{\{1,n\}}$. His G_* had the drawback that it was not unique. Using methods similar to those above we can define a unique G_*. The advantage of uniqueness is that then asymptotic fixed point theorems in the sense of Browder [1] can be proved for set-valued maps such as $G: X \to Y^{\{1,n\}}$.

Section 6: The Euler Characteristic for $M^{\underline{n}}$.

Let $\chi(M)$ = the Euler Characteristic of M. Using O'Neill's theory of a fixed point index we can prove the following.

If M is a compact closed differentiable manifold then

$$\chi(M^{\underline{n}}) = \sum_{k=1}^{n} \binom{\chi(M)}{k} \quad \text{for} \quad \chi(M) \geq 0$$

and

$$\chi(M^{\underline{2}}) = \binom{|\chi(M)|}{2} \quad \text{for} \quad \chi(M) < 0$$

Using Molski's result [3] that $M^{\underline{2}}$ is a closed 4-dimensional manifold provided M is a closed 2-dimensional manifold and several of the above calculations we can further compute for a closed compact 2-dimensional manifold M

$$H_n(M^{\underline{2}};\ Z/2) = \begin{cases} Z/2 & n = 0 \\ (Z/2)^h & n = 1 \\ (Z/2)^{\bar{\chi}+2h-2} & n = 2 \\ (Z/2)^h & n = 3 \\ (Z/2) & n = 4 \\ 0 & n \geq 5 \end{cases}$$

where $h = \dim H_1(M;\ Z) = 2 - \chi(M)$ and $\bar{\chi} = \chi(M^2)$ given above.

Section 7: Higher Homotopy Groups.

It follows quite easily from the piecewise linear approximation theorem that $\pi_q(S^{n\underline{p}}) = 0$ for $0 \leq q < n$. Adapting the "islands argument" used classically to prove $\pi_n(S^n) = Z$, we can show that $\pi_n(S^{n\underline{p}}) = 0$ for $p \geq 3$ and $\pi_n(S^{n\underline{2}}) = Z$. The induced homomorphisms mentioned above can be used to show that $\pi_n(S^{n\{1,q\}}) \neq 0$ and $\pi_n(S^{n\{2,p\}}) \neq 0$ for p prime and n even. In Section 2 we already mentioned some calculations of higher homotopy groups fro some hyperspaces of graphs. An interesting geometrical approach to calculating homotopy groups is dependent upon the following concept:

Section 8: Differentiable Set-Valued Maps.

A set-valued map $F:M \to N^S$, where M and N are differentiable manifolds and $n = \max S$, is said to be differentiable provided there is an open cover $\mathcal{U}$ of M along with a set of n differentiable maps $g_1^V, \dots, g_n^V : V \to N$ for each $V \in \mathcal{U}$ such that for each $V \in \mathcal{U}$ and $x \in V$ we have $F(x) = \{g_1^V(x), g_2^V(x), \dots, g_n^V(x)\}$ and the following compatibility condition holds: If $U, V \in \mathcal{U}$ then for each i, $1 \leq i \leq n$, there exists a j, $1 \leq j \leq n$ such that $g_i^U|U \cap V = g_j^V | U \cap V$. This definition is justified by the fact that any continuous set-valued map $F:M \to N^S$ can be approximated by a homotopic differentiable map $F': M \to N^s$. Sard's theorem can be proved for such maps and the degree can be defined by counting with appropriate weights the inverse images of a point $q \in N$ (for defining the degree we need dim M = dim N). A geometric proof of the invariance of this degree under homotopy can be given using induced "multi-submanifolds". A mod p Hopf invariant can be defined and proved to be a homotopy invariant for maps $F: S^3 \to S^{2^{\{2,3\}}}$ and this leads to the results that $\pi_3(S^{2^{\{2,3\}}}) \neq 0$.

The Pointrjagin construction (see Milnor [2]) cannot be carried out directly. But obstruction theory can be applied and the result that seem to be emerging is that $\pi_2(S^{2^{\{2,p\}}}) = Z/p \oplus \pi_2(R^{2^{\{2,p\}}})$ for p prime.

Bibliography

[1] F. E. Browder "Asymptotic fixed point theorems", Math. Ann. 185, (1970) 38-60.

[2] John W. Milnor, Topology from the Differentiable Viewpoint, The University Press of Virginia, 1965.

[3] R. Molski, On symmetric products, Fund. Math., 44 (1957), 165-170.

[4] Barrett O'Neill, "Induced homology homomorphisms for set-valued maps", Pacific J. Math. 7 (1957) 1179-1184.

[5] W. Wu, Note sur les products essentials symmetriques des espaces topologique, I Comptes Rendus des Seances de L'Academic des Science, 16, (1947), 1139-1141.

SET-VALUED FIXED POINTS THEOREMS

FOR APPROXIMATIVE RETRACTS

by

Jan W. Jaworowski*

1. The notion of approximative retract was introduced by Noguchi [9] and has recently been studied by Gmurczyk [5] and Granas [6]. The class of approximative absolute neighbourhood retracts seems very suitable for generalizing the Lefschetz fixed point theorem (see [9] and [6]). On the other hand, the Lefschetz theorem has been generalized to set-valued acyclic maps of absolute neighbourhood retracts (see [4] and [10]). The present paper contains a proof of the Lefschetz theorem for set-valued acyclic maps of compact approximative neighbourhood retracts.

2. We recall the definition of approximative retracts. Given a metric space X and a number ϵ, two maps $f,g\colon Z \to X$ are said to be ϵ-near if $d(fz,gz) < \epsilon$ for every $z \in Z$.

Let X be a compact metric space which is a subspace of a space Z and ϵ be a number. A map $r\colon Z \to X$ is said to be an ϵ-retraction if $r|X$ is ϵ-near to the identity map $1_X\colon X \to X$; X is said to be an approximative retract of Z if for every $\epsilon > 0$ there exists an ϵ-retraction $r_\epsilon\colon Z \to X$. If X is an approximative retract of any metric space Z in which it is imbedded then X is said to be an approximative absolute retract (A - AR); if, for any imbedding of X in a metric space Z, the image of X is an approximative retract of a neighbourhood of it in Z, then X is said to be an approximative absolute neighbourhood retract (A - ANR). It has been shown by Noguchi [9] that a compact metric space X is an A - AR if and only if X is homeomorphic to a closed approximative retract of the

---

* Research supported by NSF Grant GP 8928.

Hilbert cube I^∞; and X is an A - ANR if and only if it is homeomorphic to a closed subspace of I^∞, which is an approximative retract of a neighbourhood of it in I^∞.

3. A compact ANR is a retract of a space of the form $P \times I^\infty$. The following result expresses an analogous property of compact A - ANR's.

(3.1) A compact A - ANR X is homeomorphic to an approximative retract of a space of the form $P \times I^\infty$, where P is a compact polyhedron.

The proof is as in [2], Satz 2. Assuming that X is an approximative retract of a neighbourhood W in I^∞, we can construct a neighbourhood of X of the form $P \times I^\infty$ contained in W.

Homological properties of A - ANR were studied in [3] and [5]. The following results are related to [5], (7.1); [6], (3.3); and [5], (7.4). They follow easily from (3.3) below. For a proof see [8].

Throughout this section H_* will denote a continuous homology functor (e.g., the Cech homology theory) and if $f: X \to Y$ then f_* will denote the homology homomorphism $f_* = H_*(f): H_*(X) \to H_*(Y)$. It is sufficient, however, to assume that H_* is any functor from the topological category to an abelian category which satisfies the homotopy axiom and the following condition:

(3.2) If X is the intersection of a nested sequence, $X_1 \supset X_2 \supset \ldots \supset X_n \supset X_{n+1} \supset \ldots$ of closed subspaces of a compact metric space Z, then the natural map

$$H_*(X) \to \varprojlim_n H_*(X_n)$$

is a monomorphism.

(3.3) Suppose that a closed subspace X of the Hilbert cube I^∞ is an approximative retract of a neighbourhood W of X in I^∞. Then the homomorphism $i_*: H_*(X) \to H_*(W)$ induced by the inclusion $i: X \to W$ is a monomorphism.

(3.4) If X is a compact A - ANR, then there exists a number $\epsilon > 0$ such that if Z is any space and $f, g: Z \to X$ are two maps which are ϵ-near, then $f_* = g_*$.

(3.5) If a compact A - ANR is an A-retract of a space Z, then the homomorphism $i_*: H_*(X) \to H_*(Z)$ induced by the inclusion $i: X \to Z$ has a left inverse $r_*: H_*(Z) \to H_*(X)$ induced by an ϵ-retraction $r: Z \to X$.

(3.6) If X is a compact A - ANR, then the graded homology group $H_*(X)$ is of finite type.

4. Let X, Y be compact spaces and $F: X \to Y$ be a set-valued map; that is, for each $x \in X$, $F(x)$ is a non-empty subset of Y. The graph of F is $\Gamma(F) = \{(x,y) \mid y \in F(x)\} \subset X \times Y$. F is said to be continuous if $\Gamma(F)$ is closed (this is equivalent to the upper semi-continuity of F regarded as a map from X to the space of closed subsets of Y). F is said to be acyclic if each $F(x)$ is acyclic (with respect to the Čech homology theory with coefficients in a field).

Let $F: X \to Y$ be a set-valued continuous acyclic map and let $p: \Gamma(F) \to X$, $q: \Gamma(F) \to Y$ denote the projections defined by $(x,y) \to x$, $(x,y) \to y$, respectively. By the Vietoris Mapping Theorem [1], $p_*: H_*(\Gamma(F)) \to H_*(X)$ is an isomorphism. Thus F defines a homomorphism $F_* = q_* \circ p_*^{-1} : H_*(X) \to H_*(Y)$ which is said to be induced by F.

If W, X, Y and Z are compact spaces, $r: W \to X$ is a single-valued continuous map, $F: X \to Y$ is a set-valued continuous acyclic map and $i: Y \to Z$ is a topological imbedding, then the composition $i \circ F \circ r: W \to Z$ defined by $(i \circ F \circ r)x = iF(rx)$ is continuous and acyclic and

<u>(4.1)</u> $(i \circ F \circ r)_* = i_* F_* g_*$.

This is a special case of a theorem proved in [10]. The proof follows from the commutativity of the following diagram:

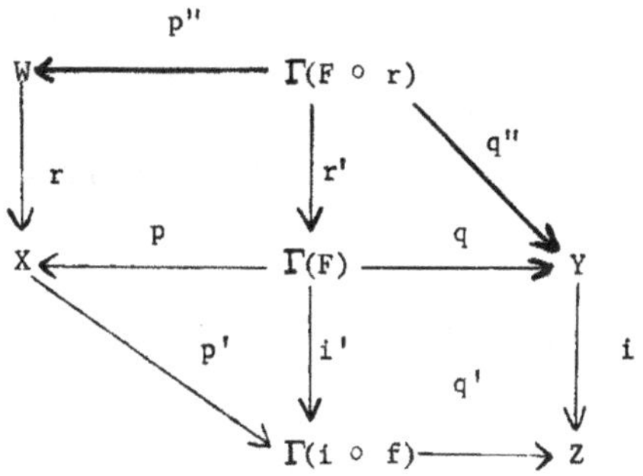

where the maps p', q', p'', q'', r' and i' are defined by $p'(x,z) = x$, $q'(x,z) = z$, $p''(w,y) = w$, $q''(w,y) = y$, $r'(w,y) = (rw, y)$, $i'(x,y) = (ix, y)$.

If $V = \{V_n\}$ is a graded vector space and $f: V \to V$ is an endomorphism of degree zero, then f is said to be a Lefschetz endomorphism if for each n the trace $tr(f_n)$ of $f_n: V_n \to V_n$ is defined in the sense of [6], (2.3) and if $tr(f_n) = 0$ for all but a finite number of n. In this case the Lefschetz number $\Lambda(f) = \sum_{-\infty < n < +\infty} (-1)^n tr(f_n)$ of f is defined. The following result follows from [7], (2.4).

(4.2) Let V and W be graded vector spaces and let $g:: V \to W$, $h: W \to V$ be graded vector space homomorphisms of degree zero such that $h \circ g: V \to V$ is a Lefschetz endomorphism. Then $g \circ h: W \to W$ is a Lefschetz endomorphism and $\Lambda(g \circ h) = \Lambda(h \circ g)$.

A set-valued continuous acyclic map $F: X \to X$ of a compact space X is said to be a Lefschetz map if the graded vector space endomorphism $F_*: H_*(X) \to H_*(X)$ is a Lefschetz endormorphism. In this case the Lefschetz number of F is defined to be $\Lambda(F) = \Lambda(F_*)$.

A point $x \in X$ is said to be a fixed point of $F: X \to X$ if $x \in F(x)$. A space X is said to be a Lefschetz space with respect to set-valued maps (or, briefly, an S-Lefschetz space) if every set-valued continuous acyclic map $F: X \to X$ is a Lefschetz map and if the condition $\Lambda(F) \neq 0$ implies that F has a fixed point. Eilenberg and Montgomery proved in [4] that every compact ANR is an S-Lefschetz space.

(4.3) Theorem. If X is a closed approximative retract of a compact S-Lefschetz space Z, then X is an S-Lefschetz space.

Proof. Suppose that $F: X \to X$ is set-valued continuous acyclic map. By (3.5) there is an ϵ-retraction $r: Z \to X$ such that $r_* \circ i_*: H_*(X) \to H_*(Z)$ is the identity, where $i: X \to Z$ is the inclusion. Consider the commutative diagram

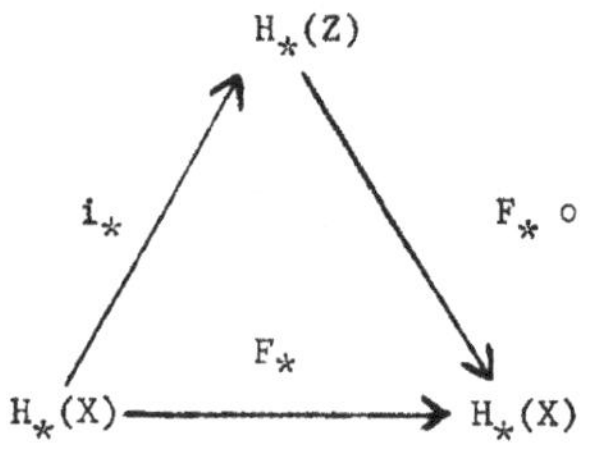

Since Z is an S-Lefschetz space and $i \circ F \circ r: Z \to Z$ is continuous and acyclic, $(i \circ F \circ r)_* = i_* \circ F_* \circ r_*$ is a Lefschetz endomrophism. It follows from (4.2) that $F_* = F_* \circ r_* \circ i_* = (F \circ r \circ i)$ is a Lefschetz endormorphism and $\Lambda(i \circ F \circ r) = \Lambda(F \circ r \circ i) = \Lambda(F)$.

Suppose that F has no fixed points. Then the graph $\Gamma(F)$ does not intersect the diagonal in $X \times X$. It follows that there exists an $\epsilon > 0$ such that if $x,y \in X$ and $d(x,y) < \epsilon$, then $y \notin F(x)$. We may assume that the retraction $r: Z \to X$ is an ϵ-retraction. It follows that the map $i \circ F \circ r: Z \to Z$ has no fixed points; for if z is a fixed point of $i \circ F \circ r$, then $z \in F(rz)$; hence $z \in X$ and $d(z,rz) < \epsilon$. This implies that $\Lambda(i \circ F \circ r) = 0$ and thus $\Lambda(F) = 0$.

(4.4) Corollary. Every compact A - ANR is an S-Lefschetz space.

This follows from (4.3), (3.1) and from the fact that $P \times I^\infty$ in (3.1) is a compact ANR and thus an S-Lefschetz space.

REFERENCES

1. E. G. Begle, The Vietoris mapping theorem for bicompact spaces , Ann. of Math. 51 (1950), 534-543.

2. K. Borsuk, Zur kombinatorischen Eigenschaften der Retrakte , Fund. Math. 21 (1933), 91-98.

3. D. G. Bourgin, Un indice dei punti uniti, Rend. Mat. Lincei 21 (1956), 395-400.

4. S. Eilenberg and D. Montgomery, Fixed point theorems for multi-valued transformations , Am. J. of Math. 58 (1946), 214-222.

5. A. Gmurczyk, On Approximative Retracts, Bull. Acad. Pol. Sci. 16 (1968), 9-14.

6. A. Granas, Fixed Point Theorems for the Approximative ANR-s , Bull. Acad. Pol. Sci. 16 (1968), 15-19.

7. J. W. Jaworowski and M. J. Powers, Λ-Spaces and Fixed Point Theorems, Fund. Math.

8. J. W. Jaworowski, Continuous homology properties of approximative retracts , (to be published).

9. H. Noguchi, A generalization of absolute neighbourhood retracts , Kodai Math. Sem. Rep., Tokyo, 1 (1953), 20-22.

10. M. J. Powers, Multi-valued mappings and fixed point theorems , thesis, Indiana University, 1968.

COMPACTNESS RELATED PROPERTIES IN HYPERSPACES

by

James Keesling

There are two somewhat related areas in the study of hyperspaces which will be dealt with in this report. They are: properties related to compactness in hyperspaces, and compactification of hyperspaces. In the first two parts of the paper we report on our work in [7] on the properties of paracompactness, meta-compactness, meta-Lindelöfness, normality, countable compactness, and pseudocompactness in hyperspaces. Surprisingly, the first three of these properties (see definitions below) are equivalent to compactness in 2^X. The last two are not. (Since the giving of the report the author has been able to show that assuming the coninuum hypothesis, normality is equivalent to compactness in 2^X. The result is announced in [8] and the details will appear in [9].)

If X is normal and T_1, then the map $F: 2^X \to 2^{\beta X}$ defined by $F(K)=Cl_{\beta X}K$ is an imbedding of 2^X onto a dense subset of $2^{\beta X}$. Thus $2^{\beta X}$ is a compactification of 2^X. In the third section of the paper we deal with the question: When is $\beta(2^X) = 2^{\beta X}$. We have been able to show that for X a well ordered countably compact space with the order topology $\beta(2^X) = 2^{\beta X}$.

By X **will always be meant** a regular T_1 topological space. The notation is basically that of [10] and [11]. By 2^X is meant the space of closed nonempty subsets of X with the finite topology [11, Definition 1.7].

1. <u>Properties Implying Compactness</u>. In [6] Ivanova showed that if $X = [0,\alpha)$ is an ordinal space with α a limit ordinal, then 2^X is nonnormal. Using this result one easily obtains the following two theorems.

<u>Theorem 1</u>: If 2^X is normal, then X is countably compact and normal.

Theorem 2: If N is the integers with discrete topology, then 2^N is not meta-Lindelöf. (See below).

Definition: A space is paracompact if each open cover has a locally finite refinement [2, p. 162]. A space is metacompact if each open cover has a point finite refinement [2, p. 229] and meta-Lindelöf (countably meta-compact in [1]) if each open cover has a point countable refinement.

The next theorem follows easily from Theorem 2.

Theorem 3: The following are equivalent:

(a) X is compact,

(b) 2^X is compact,

(c) 2^X is Lindelöf,

(d) 2^X is paracompact,

(e) 2^X is metacompact, and

(f) 2^X is meta-Lindelöf.

Another surprising application of the result of Ivanova is the next theorem.

Theorem 4: If 2^{2^X} is normal, then X is compact.

This is proved by showing that if X is not compact, then it is possible to embed a noncompact ordinal space $[0,\alpha)$ as a closed subset of 2^X. Thus $2^{[0,\alpha)}$ will be a closed subspace of 2^{2^X} and 2^{2^X} will be nonnormal.

2. Countable Compactness and Pseudocompactness. It is not necessary that 2^X be compact to be countably compact as the next theorem shows.

Definition: A space is pseudocompact if each real valued continuous function is bounded. A space is <u>strongly countably compact</u> is each countable set has compact closure.

<u>Theorem 5</u>: If X is strongly countably compact, then 2^X is pseudocompact. If in addition X is normal, then 2^X is strongly countably compact.

<u>Theorem 6</u>: If $X = [0,\alpha)$ is a countably compact ordinal space, then 2^X is strongly countably compact.

3. <u>Compactification</u>. We now turn to the problem of determining when $\beta(2^X) = 2^{\beta X}$. We will assume that X is normal and T_1 in this section so that 2^X is completely regular and the imbedding F spoken of earlier exists. The next theorem shows that the relation is possible for noncompact spaces X.

<u>Theorem 7</u>: If X is a countably compact ordinal space, then $\beta(2^X) = 2^{\beta X}$.

The proof of Theorem 7 requires the following lemma.

<u>Lemma</u>: Let $X = [0,\alpha)$ be a countably compact noncompact ordinal space and $f: 2^X \to [0,1]$ be continuous. Then there is a $\tau < \alpha$ such that if L is any closed subset of $[0,\alpha)$ and P_1 and P_2 are any nonempty closed subsets of $[\tau,\alpha)$, then $f(L \cup P_1) = f(L \cup P_2)$.

The next theorems give necessary conditions for $\beta(2^X) = 2^{\beta X}$. The proof of the first is somewhat difficult and rather unintuitive. The proof of the second is relatively straightforward.

<u>Theorem 8</u>: If $\beta(2^X) = 2^{\beta X}$, then 2^X is pseudocompact. Thus X must be pseudocompact also.

In agreement with Theorem 8 and Theorem 7, if X is a countably compact ordinal space, then 2^X is pseudocompact by Theorem 6.

<u>Theorem 9</u>: If $\beta(2^X) = 2^{\beta X}$, then 2^X is nonnormal or compact.

REFERENCES

1. G. Aquaro, "Point countable open coverings and countably compact spaces" General Topology and its Relations to Modern Analysis and Algebra II, Academic Press, 1967, pp. 39-41.

2. J. Dugundji, Topology, Allyn and Bacon, 1966.

3. Z. Frolík, "Sums of ultrafilters" Bull. Amer. Math. Soc. 73 (1967), 87-91.

4. T. Ganea, "Symmetrische Potenzen topologischer Raume" Math. Nach. 11 (1954), 305-316.

5. L. Gillman and M. Jerison, Rings of Continuous Functions, Van Nostrand, 1960.

6. V. M. Ivanova, "On the theory of the space of subsets" (in Russian) Dokl. Akad. Nauk SSSR 101 (1955), 601-603.

7. J. Keesling, "Normality and properties related to compactness in hyperspaces" Proc. Amer. Math. Soc. 24 (1970), 760-766.

8. ______________, "Normality and compactness are equivalent in hyperspaces" Bull. Amer. Math. Soc. 76 (1970), 618-619.

9. ______________, "On the equivalence of normality and compactness in hyperspaces" Pac. J. Math. (to appear).

10. K. Kuratowski, Topology I, Academic Press, 1966, p. 160-182.

11. E. Michael, "Topologies on spaces of subsets" Trans. Amer. Math. Soc. 71 (1951), 152-182.

SOME PROBLEMS CONCERNING SEMI-CONTINUOUS SET-VALUED MAPPINGS

by

K. Kuratowski

§1. Introduction. All spaces in this paper will be assumed T_1 topological.

Let F be a (set-valued) mapping assigning to each $y \in Y$ a closed subset $F(y)$ of X. Thus, $F: Y \to 2^X$, where 2^X is the space of all closed subsets of X endowed with the Vietoris (exponential) topology.

F is called _upper_ (resp. _lower_) _semi-continuous_ (briefly u.s.c. resp. l.s.c.) if the set $\{y: F(y) \cap A \neq \emptyset\}$ is closed (resp. open) in Y whenever A is closed (resp. open) in X.

Let $f: X \to Y$ be continuous. It has been shown (see e.g. [2], p. 177) that the mapping $f^{-1}: 2^Y \to 2^X$ is u.s.c. iff f is a closed mapping (symmetrically, f^{-1} is l.s.c. iff f is an open mapping).

Thus, _if X is compact, then f^{-1} is u.s.c._

The last statement will be generalized as follows.

Theorem. _Let X be compact and Y regular. Put_

(1) $$H(f,B) = f^{-1}(B),$$

where $f: X \to Y$ is continuous and $B \subset Y$ closed. Thus,

(2) $$H: Y^X \times 2^Y \to 2^X$$

(where Y^X is endowed with the compact-open topology).

Then H is u.s.c.

The proof of this theorem (which has recently found several applications) will be based on the following auxiliary theorems.

§2. Auxiliary theorems.

1. Let X be compact, Y Hausdorff and $F: Y \to 2^X$. Then F is u.s.c. iff the set $\{< x,y > : x \in F(y)\}$ is closed in $X \times Y$ (see [3], p. 58).

2. X is regular iff the set $\{< x,A > : x \in A\}$ is closed in $X \times 2^X$ (see [2], p. 167).

2a. If Y is regular and $f: X \to Y$ continuous, then the set $\{< x,B > f(x) \in B\}$ is closed in $X \times 2^Y$.

3. 2^X is Hausdorff iff X is regular (see [4], p. 163).

4. 2^X is compact iff X is compact (see [4], p. 161).

5. Let X be compact Hausdorff and $f: X \to Y$ continuous. Put $w(x,f) = f(x)$. Then $w: X \times Y^X \to Y$ is continuous (see e.g. [3], p. 77).

6. If Y is Hausdorff, then so is Y^X (see [1], p. 222).

7. If $f: X \to Y$ is continuous and $F: Y \to 2^X$ u.s.c. (l.s.c.), then the composed mapping $H = F \circ f$ is u.s.c. (l.s.c.) (see [2], p. 178).

§3. Proof of the Theorem. According to 3 and 6 the space $Y^X \times 2^Y$ is Hausdorff, and by 4, 2^X is compact.

Thus, Theorem 1 can be applied (replacing y by $< f,B >$). So we have to show that the set

$$E = \{< x,f,B >: x \in H(f,B)\}$$

is closed. Now, by (1)

$$E = \{< x,f,B >: x \in f^{-1}(B)\} = \{< x,f,B >: f(x) \in B\}$$

$$= \{< x,f,B >: w(x,f) \in B\},$$

and the last set is closed by 2a (replacing x by $< x,f >$), since w is continuous (by 5).

§4. Corollaries. Let X be compact, Y regular and $f: X \to Y$ continuous. Put

$$H_1(f,y) = f^{-1}(y) \quad \text{and} \quad H_2(f,x) = f^{-1}[f(x)].$$

Then the mappings $H_1: Y^X \times Y \to 2^X$ and $H_2: Y^X \times X \to 2^X$ are u.s.c.

Upper semi-continuity of H_1 is an obvious consequence of the Theorem. Replacing $< f,y >$ by $< f,f(x) >$ in H_1, we deduce that H_2 is u.s.c. by virtue of 7.

§5. Problems.

1. Let X be compact metric and Y metric. Then each semi-continuous mapping $F: Y \to 2^X$ is of Baire class 1 (i.e. inverse images of open sets are F_σ-sets) (see [2], p. 70).

Problem: Can compactness and metrizability be replaced by weaker assumptions?

2. Let X denote the closed interval $[0,1]$ and let for $A \subset X$, A^d denote the derivative of the set A (i.e. the set of its points of accumulation). One can easily show that A^d is a mapping of 2^X of Baire class 2 precisely (i.e. is not of class 1).

Problem: What are precise classes of derivatives of higher order?

3. Call a set-valued mapping $F: Y \to 2^X$ a u.s. α (resp. l.s. α) mapping if the set $\{y: F(y) \cap A \neq \emptyset\}$ is of multiplicative (resp. additive) Borel class α, whenever A is closed (resp. open).

Problem: Under what assumptions theorems known for $\alpha = 0$ (i.e. theorems on semi-continuous mappings) can be extended to arbitrary $\alpha < \omega_1$?

REFERENCES

1. J. L. Kelley, General topology, Van Nostrand 1955.

2. K. Kuratowski, Topology vol. I, Academic Press, 1966.

3. __________, Topology vol. II, Academic Press, 1968.

4. E. Michael, Topologies on spaces of subsets, Trans. Amer. Math. Soc. 71 (1951), pp. 152-182.

CONCERNING OPEN SELECTIONS

by

Louis F. McAuley[1]

Introduction. One of the numerous useful applications of Michael's selection theorems concerns the existence of dimension raising open mappings. I shall illustrate this as follows. Suppose that f is a completely regular monotone mapping of X onto Y where each of X and Y is a compact metric continuum. (Definitions of terms used here may be found in [2; 4; 5; 6].) Under certain conditions, we can show the existence of special types of mappings from X onto $Y \times K$ where K is also a compact metric space. Some results for special cases will appear in [5]. In these cases, each of Y and K is $[0,1] = I$. The proofs depend heavily on a selection theorem of Michael [6]. One such theorem is the following:

Theorem [5]. Suppose that f is a monotone completely regular mapping of a compact metric continuum X onto $I = [0,1]$. Furthermore, for each y in I, the space $(0_y(f), \sigma)$ of all open mappings of $f^{-1}(y)$ onto I is locally connected where $f^{-1}(y)$ is a Peano continuum. Then there is an open mapping g of a subcontinuum A of X onto $I \times I$. (Here, σ is a suitably chosen metric.)

A similar theorem is stated in [5] where open mapping is replaced by light open mapping.

There are several difficulties involved in proving theorems of this type. There are questions concerning the completeness and local connectivity (in appropriate dimensions) of the spaces $0_y(f)$ if selection theorems of Michael are to be used. Another serious difficulty involves open selections. While there may be continuous selections one from each $0_y(f)$ for each $y \in Y$, the resulting selection may not yield the desired open mapping. Each element of $0_y(f)$ is open but together they may not yield open mappings on A. We shall look at this more carefully.

---

[1] This research was supported in part by NSF Grant 6951.

A Space of Mappings. Suppose that C is a continuous collection of closed subsets of a complete metric space X, i.e., C represents a subspace of 2^X (under a Hausdorff metric).

For each $g \in C$, let $0_y(K)$ denote the space of all open mappings of g onto K, a compact metric space. Let $0(K)$ denote the collection of all $0_g(K)$, g in C, and $0(K)^*$ denote the union of the elements of $0(K)$. That is, $0(K)^*$ is the set of all open mappings f in $0_g(K)$ for each g in C.

A metric for $0(K)^*$. If $f \in 0(K)^*$, then $f \in 0_c(K)$ for some $c \in C$. Let $\hat{f}$ denote the graph of f in $X \times K$ where f is an open mapping of c onto K. For each pair of elements f, g of $0(K)^*$ where $f \in 0_a(K)$ and $g \in 0_b(K)$, let $D(f,g) = H(\hat{f},\hat{g})$ where H denotes the Hausdorff metric on the space of all closed subsets of $X \times K$ which is defined with respect to a bounded complete metric for $X \times K$ (cf. [4]). Clearly, D is a metric for $0(K)^*$ but D may not be complete.

Now, if we use continuous mappings instead of open mappings, then $(0(K)^*,D)$ is topologically complete [4]. But, we are using open mappings. If X is a Peano continuum, K is the interval $I = [0,1]$, and $C = \{f^{-1}(y) | y \in I$ where f is completely regular mapping of X onto I and $f^{-1}(y)$ is a Peano continuum$\}$, then $(0(K)^*,D)$ is also topologically complete. (Cf. [5; Theorem 3]). Furthermore, in this case, $0(K)$ is lower semi-continuous.

Question. Under what conditions is $(0(K)^*,D)$ topologically complete?

We might note that $0(K)$ is lower semi-continuous whenever C is a completely regular collection. [Cf. 4]. In fact, we can prove the following:

Theorem 1. Suppose that for each g in C, $0_g(K)$ is LC^n (in the homotopy sense) and that $0(K)$ is completely regular. Then $0(K)$ is equi-LC^n.

Theorem 2. Suppose that $(0(K)^*,D)$ is topologically complete, $0(K)$ is completely regular, for each g in C, $0_g(K)$ is LC^n, and F is a 1-1 function from $O(K)$ onto a metric space Y where $\dim Y \leq n+1$. Then for each point $y \in Y$, there is an open set U, $y \in U$, and a continuous selection φ from U into $O(K)^*$ such that $\varphi(p) \in F^{-1}(p)^*$ for each $p \in U$.

Note that we have the following application.

Theorem 3. The mapping φ induces a continuous mapping s from $F^{-1}(U)^*$ onto $U \times K$ such that $s|\varphi(p)$ is an open mapping of g onto (p,K) where $\varphi(p) \in O_g(K)$, i.e., for x in g, $s(x) = (p,\varphi(p)(x))$. (The mapping s may not be open, however.)

Question. Under what conditions is the mapping s open?

We provide one type of answer to this question. For instance, we can define a "suitable" metric σ for $O(K)^*$ such that whenever φ induces a continuous mapping s as above, s is open. The metric σ will be complete but $(O(K)^*,D)$ is not necessarily topologically equivalent to $(O(K)^*,\sigma)$.

A suitable topologically complete metric for $O(K)^*$. First, consider $(O(K)^*,D)$ to be complete. Recall that this is the case (at least, topological completeness) when $C = \{f^{-1}(y) | y \in I$ where f is a completely regular mapping of X onto I, X is a compact metric continuum, and $f^{-1}(y)$ is a Peano continuum$\}$.

Let ρ denote a metric for K. (Recall that K is compact.) For g and h in $O(K)^*$, let $\sigma(g,h) = D(g,h) + \text{g.l.b.}\{\varepsilon > 0|$ there is a $\delta > 0$ such that if $\rho(p,q) < \delta$; $p,q \in K$, then $H(g^{-1}(p),h^{-1}(q)) \leq \varepsilon\}$ where H is the Hausdorff metric on the space of all closed subsets of X defined by using a bounded complete metric for X. It is not difficult to show that σ is a complete metric for $O(K)^*$. Furthermore, using $(O(K)^*,\sigma)$ in Theorems 2 and 3, it follows that s is an open mapping.

A similar argument can be used to obtain a complete metric σ for the space $LO_g(K)$ of all light open mappings of g onto K for $g \in C$ whenever the space $(LO_g(K)^*,D)$ is topologically complete. Correspondingly, we have the complete metric space $(LO(K)^*,\sigma)$. Also, note that when $C = \{f^{-1}(y) | y \in I$ where f is a completely regular mapping of X onto I, X is a compact metric continuum, and $f^{-1}(y)$ is a Peano continuum$\}$, then the spaces $LO_g(K)$ and $LO(K)^*$ are topologically complete under the metric D defined earlier.

We will indicate here a proof of Theorem 7 of [4] which serves to illustrate the use of a selection theorem of Michael.

Theorem 7 [4]. Suppose that f is a monotone completely regular mapping of a compact metric continuum onto $I = [0,1]$. Furthermore, for each $y \in I$, the space $(O_y(f),\sigma)$ of all mappings of $f^{-1}(y)$ onto I is locally connected and also $f^{-1}(y)$ is a Peano continuum. Then there is an open mapping g of a subcontinuum A of X onto $I \times I$.

Brief indication of a proof. The collection G of the various $O_y(f)$ for $y \in I$ is (1) lower semi-continuous and (2) equi-LC^o. Furthermore, G^* (the union of the elements of G is $O(f,X)$ and with metric $\sigma(O(f,X),\sigma)$ is a complete metric space. To establish (1) and (2) above, use an argument like that in [4] referred to above.

Again, by techniques in [2] and the use of a selection theorem of Michael [Cf. 6], it follows that there is a continuous mapping φ of an open interval (a,b) of I into $O(f,K)$ such that $\varphi(x) \in O_x(f)$ for each x in (a,b). Consequently, φ induces a continuous mapping s of $f^{-1}(a,b)$ onto $(a,b) \times I$. We let $s(p) = (f(p),\varphi(f(p))(p))$ for each p in $f^{-1}(a,b)$. It follows that s is also open. Let $A = f^{-1}[c,d]$ where $a < c < d < b$. Since f is a monotone mapping, A is a continuum. And, $s|A$ is an open mapping of A onto $[c,d] \times I$. But, $[c,d] \times I \cong I \times I$. Let $h(s|A) = g$ where h is the required homeomorphism. Thus, g is the mapping claimed in the theorem.

There are generalizations of the various theorems mentioned here. For example, one of my students, Mr. David Addis, expects to include some of these in one part of his thesis.

REFERENCES

1. Anderson, R. D., A continuous curve admitting monotone open maps onto all locally connected metric continua, (ABSTRACT) BAMS Vol. 62 (1956) p. 264.

2. Dyer, Eldon and Hamstrom, M-E, Completely regular mappings, Fund. Math. Vol. 45 (1957) pp. 103-118

3. Keldys, Zero dimensional mapping of a one dimensional Peano continuum onto a square (2-cell), Doklady Akad. Nauk. SSSR (N.S.) Vol. 97 (1954), pp. 203-204 (Russian).

4. McAuley, L. F., Existence of a complete metric for a complete metric for a special mapping space, Annals of Math. Studies #60 (1966).

5. McAuley, L. F., A note on spaces of cerain non-alternating mappings onto an interval, to appear in Duke Jour.

6. Michael, E. A., Continuous Selections, I, II, III, Annals of Math, 1956-67.

7. Whyburn, G. T., Non-separated cuttings of connected point sets, TAMS, Vol. 33 (1931) pp. 444-454.

8. ____________, Non-alternating transformations, Amer. Jour. Math. Vol. 56 (1934), pp. 294-302.

9. ____________, Non-alternating interior retracting transformation, Annals of Math. (2) Vol. 40 (1939), pp. 914-921.

10. ____________, The Existence of certain transformations, Duke Math Jour. Vol. 5 (1939), pp. 647-655.

11. ____________, Analytic Topology, New York (1942) AMS Colloquium Publication, Vol. 28.

12. ____________, Monotoneity of limit mappings, Duke Math. Journ. Vol. 29 (1962) pp. 465-70.

13. Wilson, David, Monotone open and light open dimension raising mappings, Ph.D. Thesis, Rutgers University, 1969.

14. Addis, David, Strongly regular mappings and fibrations, Ph.D. Thesis, Rutgers University, 1970.

A SURVEY OF CONTINUOUS SELECTIONS

by

E. Michael

1. Introduction. Let 2^Y denote the space of all non-empty subsets of Y. If $F: X \to 2^Y$, then a selection for F is a continuous $f: X \to Y$ such that $f(x) \in F(x)$ for all $x \in X$. Our purpose is to study the existence of selections, as well as extending selections defined on a closed subset of X.

A function $F: X \to 2^Y$ is called lower semi-continuous, or l.s.c., if $\{x \in X: F(x) \cap V \neq \emptyset\}$ is open in X for every open V in Y. We will study selections only for l.s.c. functions $F: X \to 2^Y$, partly because they yield useful theorems, and partly because of the following result.

PROPOSITION 1.1. If $F: X \to 2^Y$, and if, for every $x \in X$ and every $y \in F(x)$, there exists a selection f for F (or even a selection for $F|U$ for some neighborhood U of x in X) such that $f(x) = y$, then F is l.s.c.

The following are three sources of l.s.c. maps.

EXAMPLE 1.2. Let $u: Y \to X$ be onto. Then u is open if and only if $u^{-1}: Y \to 2^X$ is l.s.c.

EXAMPLE 1.3. Let Y be a topological linear space. If $F: X \to 2^Y$ is l.s.c., so is $\Gamma \circ F$, where $\Gamma(A)$ denotes the closed convex hull of A.

EXAMPLE 1.4. Let $F: X \to 2^Y$ be l.s.c., let $A \subset X$ be closed, and let f be a selection for $F|A$. Then $G: X \to 2^Y$, defined by $G(x) = \{f(x)\}$ if $x \in A$ and $G(x) = F(x)$ if $x \notin A$, is l.s.c.

2. Selection theorems with convex sets. If Y is a topological linear space, Ⓚ(Y) denotes the space of closed, convex, non-empty subsets of Y.

THEOREM 2.1. The following properties of a T_1 - space X are equivalent.

(a) X is paracompact

(b) If Y is a Banach space, every l.s.c. $F: X \to \mathfrak{K}(Y)$ admits a selection.

The following two corollaries formally strengthen 2.1(a) → (b). They follow immediately from 2.1(a) → (b) and Example 1.3 and 1.4, respectively.

COROLLARY 2.2. If X is paracompact, Y a Banach space, and $F: X \to \tilde{\mathfrak{K}}(Y)$ l.s.c., there exists a selection for $\Gamma \circ F$.

COROLLARY 2.3. If X is paracompact, Y a Banach space, $F: X \to 2^Y$ l.s.c., and $A \subset X$ closed, then any selection for $F|A$ can be extended to a selection for F.

Another consequence of 2.1(a) → (b), using Example 1.2, is the following result of R. G. Bartle and L. M. Graves.

COROLLARY 2.4. If E is a Banach space and F a closed linear subspace of E, then the projection $E \to E/F$ admits a continuous cross-section.

We now come to an application of the equivalence of (a) and (b) in Theorem 2.1. Let us say that a space X has the _fine topology with respect to a closed covering_ $\mathfrak{B}$ of X if a subset A of X is closed whenever there exists a subcollection $\mathfrak{B}'$ of $\mathfrak{B}$ which covers A such that $A \cap B$ is closed for every $B \in \mathfrak{B}'$.

The following result of K. Morita is an easy consequence (using Zorn's lemma) of Theorem 2.1.

COROLLARY 2.5. If X has the fine topology with respect to a covering by closed paracompact subsets, then X is paracompact.

One reason for the interest of Corollary 2.5 is that it implies the following result of H. Miyazaki.

COROLLARY 2.6. Every CW-complex is paracompact.

There are several directions in which Theorem 2.1(a) → (b) can be improved. First of all, it remains valid (with the same proof) for any complete metrizable locally convex topological linear space Y. The same is, of course, true for Corollary 2.2. With considerably more effort, we can further generalize Corollary 2.2 as follows.

THEOREM 2.7. Let X be paracompact, Y a complete locally convex topological linear space, and $M \subset Y$ completely metrizable (i.e. M admits a complete metric compatible with its topology). Then, if $F: X \to 2^M$ is l.s.c., there exists a selection for $\Gamma \circ F$.

In a different direction, one can generalize Theorem 2.1(a) $\to$ (b) by letting Y be a complete metric space carrying an axiomatically defined <u>convex structure</u> which permits one to take "convex combinations" of some (but not necessarily all) ordered n-tuples of points of Y in a suitably continuous fashion. With <u>convex subset</u> of Y defined in the natural way ($A \subset Y$ is <u>convex</u> if all n-tuples in A admit convex combinations, and these always lie in A), Theorem 2.1 and Corollaries 2.2 and 2.3 remain true in this more general setting. Moreover, there is a local analogue of Corollary 2.3 in which, under suitable hypotheses, one concludes that every selection for $F|A$ can be extended to a selection for $F|U$ for some open $U \supset A$. Using these results, one can prove the following theorem, whose global part generalizes Corollary 2.4.

THEOREM 2.8. If G is a metrizable topological group, and H a complete subgroup which is (locally) isomorphic to the additive topological group of a Banach space, then the projection $G \to G/H$ admits a (local) continuous cross-section.

To prove Theorem 2.8, one defines a suitable convex structure on G in which convex combinations can be taken of all n-tuples whose elements all lie in some right translate of (a sufficiently small fixed neighborhood of the identity in) H.

3. <u>A Selection Theorem with no Linear or Convex Structure on</u> $\underline{Y}$. If X is finite dimensional, one can prove a selection theorem under purely topological

assumptions which are much weaker than anything in the previous section, and which are, in a certain sense, not only sufficient but also necessary.

DEFINITION 3.1. Let Y be a topological space.

(a) A collection $\mathcal{S}$ of subsets of Y is called <u>equi-LC^n</u> if, whenever $y \in \bigcup \mathcal{S}$ and U is a neighborhood of y in Y, there exists a neighborhood V of y in Y with the following property: If $S \in \mathcal{S}$ and $k \leq n$, every continuous map from the k-sphere into $V \cap S$ can be extended to a continuous map from the $(k + 1)$ - ball into $U \cap S$.

(b) A subset S of Y is <u>C^n</u> if $\pi_k(S) = 0$ for $k \leq n$.

In the following theorem, dim denotes the covering dimension, and $n \geq -1$.

THEOREM 3.2. Let X be paracompact, and $A \subset X$ closed with $\dim(X-A) \leq n + 1$. Let Y be a complete metric space, and $\mathcal{S}$ an equi-LC^n collection of non-empty closed subsets of Y. Let $F: X \to \mathcal{S}$ be l.s.c. Then every selection for $F|A$ can be extended to a selection for $F|U$ for some open $U \supset A$. If also every $S \in \mathcal{S}$ is C^n, then every selection for $F|A$ can be extended to a selection for F.

The global conclusion in the last sentence of Theorem 3.2 remains valid if the hypothesis that every $S \in \mathcal{S}$ is C^n is replaced by the following hypothesis: A is a weak deformation retract of X (i.e. the identity map $i: X \to X$ is homotopic over X to a retraction $r: X \to A$), every $S \in \mathcal{S}$ is compact, and F is continuous with respect to the Hausdorff metric on the compact subsets of Y (instead of merely lower semi-continuous).

REFERENCES

All results up to, but not including, Theorem 2.7 are from [1]. Theorem 2.7 is from [5]. The remainder of section 2 is from [4]. The beginning of section 3 is from [2], and the last paragraph from [3].

1. E. Michael, Continuous selection I, Ann. of Math. 63 (1956), 361-382.

2. ________ , Continuous selections II, Ann. of Math. 64 (1956), 562-580.

3. ________ , Continuous selections III, Ann. of Math. 65 (1957), 375-390.

4. ________ , Convex structures and continuous selections, Canadian J. Math. 11 (1959), 556-575.

5. ________ , A selection theorem, Proc. Amer. Math. Soc. 17 (1966), 1404-1406.

SOME COMMENTS ON THE SPACE OF SUBSETS

S. Mrówka

Western Michigan University

Given a topological space X we denote by $P(X)$ the class of all closed subsets of X and by 2^X the class of all non-empty closed subsets of X. We will be concerned with both the convergence and the topology in $P(X)$.

We consider the convergence of generalized sequences (nets). According to the definition, a generalized subsequence can have a larger set of indices than the original sequence. It is frequently useful to have an estimation for the size of the involved sets of indices. Given a cardinal $\underline{m}$, we denote by $D_{\underline{m}}$ the set of all finite subsets of a set of cardinality $\underline{m}$; $D_{\underline{m}}$ is directed by inclusion. A generalized sequence having $D_{\underline{m}}$ as the set of indices will be called an <u>$\underline{m}$-sequence</u>. It is known that all the convergence phenomena can be adequately discussed in terms of $\underline{m}$-sequences (reason: $D_{\underline{m}}$ can be mapped in an isotonic way onto any directed set of cardinality $\leq \underline{m}$) and $D_{\underline{m}}$ can provide the desired estimation.

Convergence of generalized sequences $\{A_n : n \in D\}$ of members of $P(X)$ is defined via limit superior (Ls) and limit inferior (Li):

(1) $p \in \mathrm{Ls}\{A_n : n \in D\} \equiv$ for every neighborhood U of p, $U \cap A_n \neq \emptyset$ for arbitrary large $n \in D$;

(2) $p \in \mathrm{Li}\{A_n : n \in D\} \equiv$ for every neighborhood U of p, $U \cap A_n \neq \emptyset$ for almost all $n \in D$.

If $\mathrm{Ls}\{A_n : n \in D\} = \mathrm{Li}\{A_n : n \in D\}$, then the generalized sequence $\{A_n : n \in D\}$ is called convergent (to the common value of Ls and Li, which is denoted by $\mathrm{Lim}\,\{A_n : n \in D\}$). For more details on these definitions the reader is referred to [4].

In [3] we have announced the following

Theorem 1. <u>For an arbitrary space</u> X, $P(X)$ <u>is compact relative to the convergence</u> (<u>i.e.</u>, <u>every generalized sequence of elements of</u> $P(X)$ <u>contains a</u>

<u>convergent generalized subsequence</u>).

Inasmuch as the proof of this theorem was only hinted in [3] we shall take this opportunity to present the details.

As it was pointed out in [3], the Tihonov theorem applied to the product of two-point discrete spaces yields

The Diagonal Theorem. <u>If</u> $\{x_n^{(\alpha)}:n \in D\}$, $\alpha \in A$, <u>is a collection of generalized sequences of</u> 0's <u>and</u> 1's , <u>then there exists a generalized sequence</u> $\{n_k:k \in E\}$ <u>of elements of</u> D <u>such that, for every</u> $\alpha \in A$, $\{x_{n_k}^{(\alpha)}:k \in E\}$ <u>is a convergent generalized subsequence of</u> $\{x_n^{(\alpha)}:n \in D\}$.

Now, a generalized sequence $\{A_n:n \in D\}$ is convergent iff $\mathrm{Ls}\{A_n:n \in D\} \subset \mathrm{Li}\{A_n:n \in D\}$. Translating this inclusion we see that $\{A_n:n \in D\}$ is convergent iff

(3) <u>for every open set</u> U , <u>if</u> $U \cap A_n \neq \emptyset$ <u>for arbitrary large</u> $n \in D$, <u>then</u> $U \cap A_n \neq \emptyset$ <u>for almost all</u> $n \in D$.

Given a generalized sequence $\{A_n:n \in D\}$ we consider the collection $\{x_n^{(U)}:n \in D\}$ of generalized sequences of 0's and 1's defined by $x_n^{(U)} = 1$ if $U \cap A_n \neq \emptyset$ and $x_n^{(U)} = 0$, if otherwise. Applying the Diagonal Theorem, we infer by (3), that the corresponding generalized subsequence $\{A_{n_k}:k \in E\}$ is convergent. Theorem 1 is shown.

Proofs of Theorem 1 which do not depend on the Tihonov Theorem can be found in the literature. Usually, however, they involve certain separation axiom on the underlying space. See, for instance, [2].

Inasmuch as the Diagonal Theorem refers to convergence in the products, it is easy to obtain an estimation for the set E of indices of the subsequence. In fact, if $D = D_{\underline{m}}$ and card $A \leq \underline{m}$, then as E we can also take the set $D_{\underline{m}}$. On the other hand, if $\underline{B}$ is a base in X , then the convergence of $\{A_n:n \in D\}$ is equivalent to condition (3) with U restricted to be a member of $\underline{B}$. These two facts combined yield an estimation for the size of the subsequence in Theorem 1.

Theorem 1a. <u>If</u> X <u>has a base of cardinality</u> $\leq m$, <u>then every</u> <u>m-sequence</u>

of points of P(X) contains a convergent m-subsequence.

For $m = \aleph_0$ Theorem 1a yields the classical result of Hausdorff, Urysohn and Kuratowski: P(X) *is sequentially compact whenever* X *is second countable*. Sierpinski has shown (assuming the continuum hypothesis) that in metric spaces the converse of the above holds. Precisely

Theorem 2 (Sierpinski) *If* X *is a metric space and* P(X) *is sequentially compact, then* X *is second countable*.

Recently, a student of mine, Mr. Chimenti, has shown that Theorem 2 fails for non-metrizable spaces. He shows (among others) that sequential compactness of P(X) is implied by the *descending chain condition* (= every tranfinite strictly decreasing sequence of closed subsets of X is at most countable). It is not known, however, if sequential compactness of P(X) is equivalent to one of the commonly known "countability conditions" on X.

Let us now turn to questions concerning topology in P(X). All topologies will be assumed to be Hausdorff. One of the topologies in P(X) is the so-called Vietoris topology; by definition, a subbase for this topology consists of all sets having one of the following forms: $\{F \in P(X): F \cap G \neq \emptyset\}$ and $\{F \in P(X): F \subset G\}$, where G is an arbitrary open subset of X. Note that in this topology the empty set is an isolated point of P(X); in other words, 2^X is closed in P(X).

It has been shown in [4] that the Vietoris topology agrees with the above-discussed convergence of generalized sequences in P (X) if and only if X is compact. If X is locally compact, then the convergence in P(X) can still be topologized; the topology in question - called in [4] the lbc-*topology* - has a similar subbase as the Vietoris topology with sets of the second form being replaced by those of the form $\{F \in P(X): F \cap \overline{V} = \emptyset\}$, where V is an arbitrary open subset of X with compact closure. No further improvement of this result is possible; in fact, we have shown in [4] that

Theorem 3. *If the convergence in* P(X) *can be topologized then* X *is locally compact*.

The above theorem can be shown in a much easier way than it was done in [4]; namely, it is a rather straighforward consequence of Theorem 1. Indeed, if $P(X)$ can be topologized then $P(X)$ is compact. On the hand, the class $P_1(X)$ of all subsets of X having at most one point is closed in $P(X)$ (if A_n, $n \in D$, contain at most one point, then $\mathrm{Li}\{A_n : n \in D\}$ contains at most one point); hence $P_1(X)$ is compact. Finally, X is homeomorphic to the open subset $P_1(X)\setminus\{\emptyset\}$ of $P_1(X)$ (reason: $\lim\{x_n : n \in D\} = x$ iff $\mathrm{Lim}\{\{x_n\} : n \in D\} = \{x\}$); consequently, X is locally compact.

In [1] Fell considered a variation of the lbc-topology; namely, he replaced $\overline{V}$ in $\{F \in P(X) : F \cap \overline{V} = \emptyset\}$ by an arbitrary compact subset C of X. However, this topology agrees with the lbc-topology iff X is locally compact.

Note that all three of the above discussed topologies can be obtained as particular cases of a general schema. Let $\underline{R}$ and $\underline{S}$ be classes of subsets of X; by the $\underline{R}$-$\underline{S}$-topology in $P(X)$ we shall mean the topology having as a subbase all sets having one of the forms $\{F \in P(X) : F \cap G \neq \emptyset\}$ and $\{F \cap P(X) : F \cap S = \emptyset\}$, where $G \in \underline{R}$, $S \in \underline{S}$. The Vietoris topology, the lbc-topology, and the Fell topology coincide with the R-S-topology, where R is the class of all open subsets of X and S is, respectively, the class of all closed subsets, the class of all compact closures of open subsets, and the class of all compact subsets. It is natural to ask when two pairs of classes yield the same topology. More generally we can pose this question relative to a subclass $\underline{F} \subset P(X)$. The interesting part of this question refers to the topology defined by the subbase consisting only of sets of the second of the above mentioned forms. This leads us to the following concept: we shall say that a class $\underline{S}$ is $\underline{F}$-finer than $\underline{S}'$ provided that for every $F \in \underline{F}$ and for every $S \in \underline{S}'$ with $F \cap S = \emptyset$ there are $S_1, \ldots, S_n \in \underline{S}$ such that $F \cap S_i = \emptyset$ and $S \subset S_1 \cup \ldots \cup S_n$. We say that $\underline{S}$ and $\underline{S}'$ are $\underline{F}$-equivalent provided that $\underline{S}$ is $\underline{F}$-finer than $\underline{S}'$ and vice versa. It turns out that these concepts provide a uniform way of dealing with various topological notions; we plan to publish these details in the near future.

REFERENCES

[1] J. M. G. Fell, A Hausdorff topology for the closed subsets of a locally compact non-Hausdorff space, Proc. Amer. Math. soc. 13 (1962), 472-476.

[2] Z. Frolik, Concerning topological convergence of sets, Czech. Math. J. 10 (85) (1960), 168-180).

[3] S. Mrówka, On almost-metric spaces, Bull. Acad. Polon. Sci., Cl. III, 5 (1957), 123-127.

[4] ________, On the convergence of nets of sets, Fund. Math. 45 (1958), 237-246.

SOME RESULTS ON MULTI-VALUED CONTRACTION MAPPINGS

by

Sam B. Nadler, Jr.

1. Introduction. If (X, d) is a metric space, then

(a) $\underline{2^X} = \{C \mid C$ is a nonempty compact subset of $X\}$,

(b) $\underline{CL(X)} = \{C \mid C$ is a nonempty closed subset of $X\}$,

(c) $\underline{N(\epsilon, C)} = \{x \in X \mid d(x,c) < \epsilon$ for some $c \in C\}$ if $\epsilon > 0$ and $C \in CL(X)$,

and

(d) $$\underline{H(A,B)} = \begin{cases} \inf \{\epsilon > 0 \mid A \subset N(\epsilon,B) \text{ and } B \subset N(\epsilon,A)\}, & \text{if the infimum exists} \\ \infty, & \text{otherwise} \end{cases}$$

when $A, B \in CL(X)$.

The pair $(CL(X), H)$ is a generalized metric space in the sense of Luxemburg [2] and H is called the generalized Hausdorff distance induced by d (in general H depends on d, but we shall not notate this dependency except where confusion may arise).

Let (X, d) be a metric space. A function $F: X \to CL(X)$ is called a multi-valued contraction mapping (abbreviated m.v.c.m.) iff there exists a fixed real number $\lambda < 1$ such that $H(F(x), F(y)) \leq \lambda \cdot d(x, y)$ for all $x, y \in X$.

Some results in this paper were presented to the American Mathematical Society in November, 1967; an abstract of that talk appears in [4]. A slightly different version of Theorem 5 of [5] was announced later in [3]. The notion of a multi-valued contraction mapping does not seem to have been considered prior to the announcement in [4].

In this paper we sketch briefly fixed point theorems the author has obtained for both global and local multi-valued contraction mappings. We also answer a question about local contractions posed in [5]. Many of these results appear in a less general setting in [5], and more general versions of some of these results are being prepared for publication by H. Covitz and myself (see 3 below).

We point out that in [5] the image of a point was in general required to be nonempty, closed, and bounded. The boundedness restriction was imposed so that the hyperspace was a genuine metric space. It is not necessary that the hyperspace be metric and in many of the proofs in [5] the boundedness of each point image was not used. Hence, many of these proofs can be done in the more general and natural settings of mappings into CL(X) without essential alteration.

2. Fixed Point Theorems. The first theorem below can be proved by the iteration procedure used in the proof of Theorem 5 of [5]. The boundedness of point images was not used in that proof, the curcial fact being that $H(F(x), F(y)) < \infty$.

Theorem 1. Let (X, d) be a complete metric space. If $F: X \to CL(X)$ is a m.v.c.m., then F has a fixed point (i.e., there is a $p \in X$ such that $p \in F(p)$).

In [1] Edelstein defined the notion of an (ϵ, λ) - uniformly locally contractive (single-valued) mapping. This notion was extended in [5] to set-valued mappings. We say that a function $F: X \to CL(X)$ is an (ϵ, λ) - uniformly locally contractive multi-valued mapping (where $\epsilon > 0$ and $0 \leq \lambda < 1$) provided that $H(F(x), F(y)) \leq \lambda d(x,y)$ for all $x, y \in X$ such that $d(x, y) < \epsilon$. Theorem 6 of [5] stated that these types of local contractions on complete ϵ-chainable spaces have fixed points if each point image is a nonempty compact set. This result was used to obtain, via the inverse function, fixed point theorems for locally expansive single-valued mappings which were not necessarily one-to-one. These theorems corrected and extended a result of Edelstein [1]. However, due to the compactness requirement on point images in Theorem 6 of [5], we needed a compactness requirement on the inverse images of points under a single-valued locally expansive mapping. Whether or not Theorem 6 was valid for (ϵ, λ)-uniformly locally contractive multi-valued mappings whose point images are nonempty, closed, and bounded was left unanswered (see the remark at the end of Section 3 of [5]). We answer this question in the next theorem and, in fact, in the more general setting of nonempty closed (not necessarily bounded) set-valued mappings. The absence of the boundedness requirement here is especially useful because, for continuous locally expansive mappings defined on a closed subspace A of X, the inverse of a point is always closed in X. Thus, no topological conditions need to be placed on the inverse images of points (compare with Theorem 7 of [5]). Furthermore, Theorem 8 of [5] is now an immediate corollary of Theorem 3 below.

Theorem 2. Let (X, d) be a complete ε-chainable metric space. If $F: X \to CL(X)$ is an (ε, λ)-uniformly locally contractive multi-valued mapping, then F has a fixed point.

The proof parallels the proof of Theorem 6 in [5] using the fact that boundedness of point images was not important in Theorem 5 of [5]. We sketch the proof here.

Proof. Let $d_\varepsilon: X \times X \to [0, \infty)$ be given by

$$d_\varepsilon(x, y) = \inf\{\sum_{i=1}^{n} d(x_{i-1}, x_i) \mid x_0 = x, x_1, \ldots, x_n = y \text{ is an } \varepsilon\text{-chain from } x \text{ to } y\}$$

for each $(x, y) \in X \times X$. It is easy to verify that (X, d_ε) is a complete metric space. Let H_ε be the generalized Hausdorff distance induced by d_ε. Note that, since $d(x, y) < \varepsilon$ implies $d(x, y) = d_\varepsilon(x, y)$, $CL(X)$ with respect to d is the same set as $CL(X)$ with respect to d_ε. It follows, as in the proof of Theorem 6 of [5], that F is a m.v.c.m. with respect to d_ε and H_ε. Thus, by Theorem 1, F has a fixed point.

For definitions of terms used in the next theorem see [5].

Theorem 3. Let (X, d) be a complete ε-chainable (respectively, well-chained) metric space, let A be a nonempty subset of X, and let $f: A \to X$ be an (ε, λ)-uniformly locally expansive (continuous) mapping of A onto X. If $f^{-1}(x)$ is closed in X for each $x \in X$ and $f^{-1}: X \to CL(X)$ is ε-nonexpansive (respectively, uniformly ϵ-continuous), then f has a fixed point.

Corollary 6.1 of [1] states "If f is a one-to-one (ϵ, λ)-uniformly locally expansive mapping of a metric space Y onto an ϵ-chainable complete metric space $X \supset Y$ then there exists a unique ξ such that $f(\xi) = \xi$. The proof given is that f^{-1} is (ϵ, β)-uniformly locally contractive for some $\beta < 1$. The following example shows that this is not necessarily the case and that Corollary 6.1 is false.

Example. Let $S = \{1, 2, 3, 4\}$ with absolute value distance. Define $f: S \to S$ by

$$f(x) = \begin{cases} 2, & x = 1 , \\ 4, & x = 2 , \\ 1, & x = 3 , \\ 3, & x = 4 . \end{cases}$$

Letting $\epsilon = 1\frac{1}{4}$ and $\lambda = 1\frac{1}{4}$ we see that all of the hypotheses in Corollary 6.1 are satisfies. However, f has no fixed point.

Remark. Theorem 3 corrects Corollary 6.1 of [1] and also extends it by not requiring f to be one-to-one.

Remark. A problem. Let (X, d) be a complete metric space and let $F: X \to 2^X$ be a m.v.c.m. Define $\hat{F}: 2^X \to 2^X$ by $\hat{F}(A) = \cup\{F(a) | a \in A\}$. It is easy to prove [5] that $\hat{F}$ is a contraction mapping (in the sense of Banach) of 2^X into 2^X (with the Hausdorff distance H induced by d). Since (X, d) is complete, so is $(2^X, H)$ and therefore, using the classical result of Banach, $\hat{F}$ has a fixed point in 2^X. We call this fixed point of $\hat{F}$ the invariant set of F, denoted by $\mathcal{J}(F)$. Let $\varphi(F) = \{x \in X | x \in F(x)\}$, i.e., $\varphi(F)$ is the fixed point set of F. We know from Theorem 1 that $\varphi(F) \neq \emptyset$. Furthermore, an application of the Banach iteration procedure applied to $\hat{F}$ at a point $p \in \varphi(F)$ gives that

$\varphi(F) \subset \mathcal{J}(F)$. It would be interesting to know under various circumstances how fixed point sets sit in invariant sets of multi-valued contraction mappings. For example, suppose X is a closed convex subset of a Banach space with norm $\| \ \|$. Let $F: X \to 2^X$ be a m.v.c.m. If $\overline{co}(A)$ denotes the intersection of all closed convex sets containing $A \in 2^X$, then $\overline{co}$ can be thought of as a function from 2^X into 2^X. In [5] it was shown that $\overline{co} : 2^X \to 2^X$ is nonexpansive. Hence, $\overline{co}F : X \to 2^X$ given by $(\overline{co}F)(x) = \overline{co}(F(x))$ for each $x \in X$ is also a m.v.c.m. How are the sets $\varphi(F)$, $\varphi(\overline{co}F)$, $\overline{co}(\varphi(F))$, $\overline{co}(\varphi(\overline{co}F))$, $\overline{co}\mathcal{J}(F)$, $\mathcal{J}(\overline{co}F)$, etc. interrelated?

3. Added in proof. The joint paper by Professor Covitz and myself will appear in the Israel Journal under the title "Multi-valued contraction mappings in generalized metric spaces."

Bibliography

1. M. Edelstein, "An extension of Banach's contraction principle," *Proc. Amer. Math. Soc.*, 12 (1961), pp. 7-10.

2. W. A. J. Luxemburg, "On the convergence of successive approximations in the theory of ordinary differential equations, II," *Koninkl. Nederl. Akademie van Wetenschappen*, Amsterdam, *Proc.* Ser. A (5) 61, and *Indag. Math.* (5), 20 (1958), pp. 540-546.

3. J. T. Markin, "Fixed point theorems for set valued contractions", *Notices Amer. Math. Soc.*, 15 (1968), p. 373.

4. S. B. Nadler, Jr., "Multi-valued contraction mappings," *Notices Amer. Math. Soc.*, 14 (1967), p. 930.

5. S. B. Nadler, Jr., "Multi-valued contraction mappings," *Pacific J. of Math.*, 30 (1969), pp. 475-487.

ON THE KAKUTANI FIXED POINT THEOREM AND ITS RELATIONSHIP WITH THE SELECTION PROBLEM I.

by

SURENDRA-NATH PATNAIK

1. Introduction. Kakutani [1] has proved the following theorem (which is a generalization of the well-known Brouwer's fixed point theorem which asserts that every single-valued continuous function from a closed n-simplex into itself has a fixed point.):

Theorem. (Kakutani). Let M be a compact convex subset of the Euclidean n-space and let T be an upper semi-continuous multiple-valued function from M into itself such that for every $x \in M$, the image set $T(x)$ is convex. Then T admits a fixed point, i.e., a point x_0 with $x_0 \in T(x)$.

Definition 1.1. A multiple-valued function $F: X \to Y$ is a law assigning to each point x of a topological space X, a closed non-empty subset $F(x)$ of a topological space Y. A multiple-valued (m. v.) function is called upper-semi-continuous (u. s. c.) if for all $x_0 \in X$ and for each neighbourhood N of $T(x_0)$ there exists a neighbourhood M of x_0 satisfying $T(M) \subset N$.

In the present paper, we obtain some small extensions of Kakutani's theorem and begin a study of its relationship with the selection problem [2]. We will discuss in a subsequent paper in greater detail and depth, the inter-relationship with convexity, the selection problem and fixed point theorems for set-valued maps for the category of spaces admitting simplicial decompositions, i.e., the category of compact polyhedra, for example.

2. Let $X^{(n)}$ denote the nth barycentic subdivision of the closed n-simplex X and let $T: X \to X$ be a u. s. c., m.v., function from X into itself. For each vertex ${}^n x$ of $X^{(n)}$, we choose an arbitrary point ${}^n y$ of the

non-empty closed set $T({}^n x)$. This correspondence which associates to each vertex ${}^n x$ of $X^{(n)}$, a definite point ${}^n y$ of $X^{(n)}$, when extended linearly inside each simplex $X^{(n)}$, defines a single-valued continuous function ${}^{(n)}t$ from X into itself. Since X is a closed n-simplex, by the Brouwer fixed point theorem, there exists a point ${}^n x \in X$ with ${}^{(n)}t({}^n x) = {}^n x$. This being true for the successive barycentric subdivisions, first, second, etc., we have a sequence of points $\{{}^n x\}$ and single-valued continuous functions $\{{}^{(n)}t\}$ satisfying ${}^{(n)}t\,({}^n x) = {}^n x$, for $n = 1, 2, \dots$. Since the space X is compact, there exists a subsequence $\{{}^{n_\alpha} x\}$ of $\{{}^n x\}$ which converges to a point $x_o \in X$. Let this point x be inside some m-simplex Z whose vertices are $z_1, z_2, \dots, z_{m+1}$ of the barycentric subdivision ${}^{(n)}X$. Hence,

$$
{}^n x \longrightarrow x_o, \qquad {}^n z_i \in F({}^n x_i)
$$

$$
{}^n z_i \longrightarrow \bar{z}_i, \qquad i = 1, 2, \dots, m+1,
$$

This implies by the upper semi continuity of F, $\bar{y}_i \in T(x)$, for $i = 1, 2, \dots, m+1$.

Definition 1.2. $d({}^n t(x), T(x)) = \text{g.l.b.}\,\{d(z_n, z) \mid z \in T(x),\ z_n = t(x)\}$.

Definition 1.3. A sequence of single-valued functions $\{f_n\}$ is uniformly convergent to a limit function f if for each $\epsilon > 0$, there exists an integer N such that for all $n \geq N$ and for each $x \in X$, we have, $|f_n(x) - f(x)| < \epsilon$.

A simple result in calculus states that the limit of a uniformly convergent sequence of continuous functions is continuous.

Definition 1.4. A single-valued continuous function $\bar{t}$ is called a selection for the u. s. c., m.v., function T if $\bar{t}(x) \in T(x)$ for all $x \in X$. [2] (Here, the functions $\bar{t}$ and T are from the same space into itself.)

3. Proposition 2.1. Let ${}^{n}t\colon X \to X$ be defined as the continuous single-valued function which assigns to each vertex ${}^{n}x$ of the nth barycentric subdivision ${}^{(n)}X$, a definite point (arbitrarily chosen) ${}^{n}y \in T({}^{n}x)$ where T is u.s.c., m.v., function from X into itself. Then given a $\epsilon > 0$, there exists an integer N such that for all $n \geq N$, $d({}^{(n)}t(x), T(x)) < \epsilon$.

Theorem 2.2. Let the sequence of single-valued functions $\{{}^{n}t\}$ (as defined in above) converge uniformly to a single-valued function $\bar{t}$. Then, this $\bar{t}$ is a selection for the m.v., u.s.c., function T.

Proof. From the definition of the functions $\{{}^{n}t\}$ and the uniform convergence of the functions $\{{}^{n}t(x)\}$ to $t(x)$, we have,

$$d(t(x), T(x)) \leq d(t(x), {}^{n}t(x)) + d({}^{n}t(x), T(x)) \leq \epsilon + \epsilon = 2\epsilon$$

for $n > N(\epsilon)$.

Theorem 2.3. Let T be an u.s.c., m.v., function from a closed n-simplex X into itself. Let the sequence of single-valued continuous functions $\{{}^{(n)}t\}$ (as defined before) converge uniformly. Then there exists a fixed point under T.

Proof. From the above theorem and the Brouwer fixed point theorem, there exists points ${}^{n}x$ satisfying ${}^{(n)}t\,({}^{n}x) = {}^{n}x$, for $n = 1, 2, 3, \ldots$.

This sequence of points $\{{}^{n}x\}$ converge to a point x_o satisfying the condition $x_o \in T(x_o)$ as the sequence of functions $\{{}^{(n)}t\}$ converge to the function $\bar{t}$ of the previous theorem.

REFERENCES

1. S. Kakutani, A generalization of Brouwer's fixed point theorem, Duke Math. J. 8 (1941), pp. 457-459.

2. E. Michael, Continuous selections I, Ann. of Math. 63 (1956), pp. 361-382.

3. S. Patnaik, Ph.D., Thesis, University of Illinois, Urbana, Illinois, 1963.

LEFSCHETZ FIXED POINT THEOREMS FOR MULTI-VALUED MAPS

by

M. J. Powers

1. Introduction. A Lefschetz space is a space X such that for every continuous map $f: X \to X$, the Lefschetz number $\Lambda(f) = \sum_k (-1)^k tr(f_{*k})$ is well-defined and when $\Lambda(f) \neq 0$, f has a fixed point. (f_{*k}) is the induced homomorphism w.r.t. a given homology theory. It is well known, for example, that compact ANR's are Lefschetz spaces. In particular, the homology of a Lefschetz space is of finite type.

Since the development of the Leray- Schauder fixed point theory, there has been interest in considering non-compact spaces and compact self-mappings of these spaces. The purpose here is to briefly sketch one method of generating these "generalized Lefschetz spaces" for multi-valued maps. The theory appears in greater detail and generality in [5] and will be published elsewhere [6]. The theory in the single-valued setting may be found in [4].

2. If X, Y are topological spaces and $F: X \to Y$ is a multi-valued map, by upper semi-continuity (u.s.c.) for F we understand that $F(x)$ is a compact subset of Y for each x in X and $\{x: F(x) \subseteq U\}$ is open in X for each open set U in Y. The graph of F is denoted $\Gamma(F) = \{(x,y): y \in F(x)\}$. The two natural projections are denoted $p: \Gamma(F) \to X$ and $q: \Gamma(F) \to X$ and $q: \Gamma(F) \to Y$. Finally, multi-valued maps are always denoted by capital letters and single-valued maps by lower case letters.

3. The first problem in considering Lefschetz theorems for multi-valued maps is in defining an induced homomorphism w.r.t. a certain homology functor. Several approaches to this problem have been considered. The most successful approach by Eilenberg and Montgomery [2] employs the Vietoris Mapping theorem [1].

Let $\check{H}$ denote the Čech homology functor with coefficients in the field Q of rational numbers. Let X,Y be compact, T_2-spaces and $f: X \to Y$ a continuous map which is inverse acyclic (i.e., $f^{-1}(y)$ is acyclic w.r.t. $\check{H}$ for each $y \in Y$). Then according to the Vietoris theorem, f induces an isomorphism of homology groups in each dimension.

If $F: X \to Y$ is u.s.c. and acyclic (i.e., F(x) is an acyclic set for each $x \in X$), then $p: \Gamma(F) \to X$ is inverse acyclic and $\check{H}(F) = F_*$ can be defined by the composition

$$\check{H}(X) \xrightarrow{p_*^{-1}} \check{H}(\Gamma(F)) \xrightarrow{q_*} \check{H}(Y).$$

An extension of the Vietoris theorem is necessary if non-compact spaces are to be used. This extension question is interesting in its own right; but for the purposes here, it is enough to consider the functor $\vec{H}$ defined below.

Let S denote the category of T_2-spaces and continuous maps. For X in S, let $K_X = \{K: K \subseteq X,\ K \text{ compact}\}$. Then $\vec{H}(X)$ is defined by taking a direct limit:

$$\vec{H}(X) = \text{dir lim } \{\check{H}(K)\}_{K_X} \quad .$$

For f in S, $\vec{H}(f)$ is defined in the natural way and we have $\vec{H}: S \to \mathcal{A}$, where $\mathcal{A}$ is the category of graded vector spaces over Q and homomorphisms of degree zero. It can be shown that $\vec{H}$ is a functor satisfying the homotopy axiom.

This functor gives rise to a Vietoris theorem for spaces in S and maps f which are continuous, inverse acyclic, and <u>proper</u> (i.e., the inverse image of compact sets is compact). Then for X,Y in S and $F: X \to Y$ u.s.c. and acyclic, $p: \Gamma(F) \to X$ is acyclic and proper. Hence, $\vec{H}(F) = F_*$ can be defined as before: $\vec{H}(F) = \vec{H}(q) \circ \vec{H}(p)^{-1}$. An elementary but extremely useful result can be noted here.

(3.1) <u>Theorem.</u> Let $F,G: X \to Y$ be u.s.c. acyclic maps where X,Y are in S. If either of the following conditions hold, then $F_* = G_*$.

(i) G is a cross-section of F (i.e., $G(x) \subseteq F(x)$ for each $x \in X$.

(ii) F is homotopic to G under an u.s.c. acyclic homotopy.

4. The composition of two multi-valued maps $F: X \to Y$ and $G: Y \to Z$ is defined by $(G \circ F)(x) = \bigcup_{y \in F(x)} G(y)$. If F and G are continuous, then $G \circ F$ is also continuous. However, the property of being acyclic does not carry over. For the purposes of this note it is only necessary to compose maps when one is single-valued. Then $G \circ f$ is acyclic when G is acyclic and $g \circ F$ is acyclic and g is <u>acyclic on sets</u> (i.e., g sends acyclic sets to acyclic sets).

(4.1) <u>Theorem</u>. Let $F: X \to Y$ and $G: Y \to Z$ be u.s.c. and acyclic. Then if $(G \circ F)$ is acyclic, $(G \circ F)_* = G_* \circ F_*$.

5. We now turn to the definition of the Lefschetz number. Details can be found in [4].

(5.1) <u>Definition</u>. A homomorphism $f: V \to V$ in $\mathcal{Q}$ has <u>finite type</u> if

(i) in each dimension k, $(f_k)^n (V_k)$ is finite dimensional for some positive integer n, and

(ii) in all but a finite number of dimensions, $(f_k)^n (V_k) = 0$ for some integer n.

If $f: V \to V$ is of finite type, the <u>trace</u> of $f_k: V_k \to V_k$ can be defined by $tr(f_k) = tr(f_k | (f_k)^n(V_k))$, where n is chosen so that $(f_k)^n(V_k)$ is finite dimensional. It is easy to show that this definition does not depend on n.

(5.2) <u>Definition</u>. If f is of finite type, the <u>Lefschetz number</u> of f is $\Lambda(f) = \sum_k (-1)^k tr(f_k)$.

The following lemma is crucial to the proof of (6.2) and is proved by reducing to the known result in the finite dimensional case.

(5.3) Lemma. Let $f: V \to W$ and $g: W \to V$ be homomorphisms in $\mathcal{Q}$. If $g \circ f$ has finite type, then $f \circ g$ also has finite type and $tr(g \circ f) = tr(f \circ g)$ in each dimension.

(5.4) Definition. Let $X, Y \in S$ and let $F: X \to Y$ be continuous and acyclic. Then if F_* is of finite type, the Lefschetz number of F is defined to be $\Lambda(F) = \Lambda(F_*)$. F is a Lefschetz map if F_* is of finite type and if F has a fixed point when $\Lambda(F) \neq 0$.

6. We can now define the desired generalization of a Lefschetz space. All maps are understood to be u.s.c.

(6.1) Definition. A space X in S is an $M\Lambda$-space if every compact acyclic multi-valued map $F: X \to X$ is a Lefschetz map. A space X is a Λ-space if every compact map $f: X \to X$ is a Lefschetz map. (F is compact if $\overline{F(X)}$ is compact.)

It is possible to generate Lefschetz maps and new $M\Lambda$-spaces by factoring through known $M\Lambda$-spaces. The following result is typical of this type of theorem.

(6.2) Theorem. Let X be a T_3-space and suppose that for each open cover α of X there is an $M\Lambda$- space Y_α and continuous maps $g_\alpha: X \to Y_\alpha$ and $h_\alpha: Y_\alpha \to X$ such that

(i) g_α is acyclic on sets,

(ii) $h_\alpha \circ g_\alpha$ is homotopic to 1_X,

(iii) $h_\alpha \circ g_\alpha$ and 1_X are α-near (i.e., for each $x \in X$, there is an element U of α containing both $h_\alpha \circ g_\alpha(x)$ and x).

Then X is an $M\Lambda$- space.

We will sketch the idea of the proof. Let $F: X \to X$ be compact and acyclic and consider any open cover α of X. Then we have the diagram

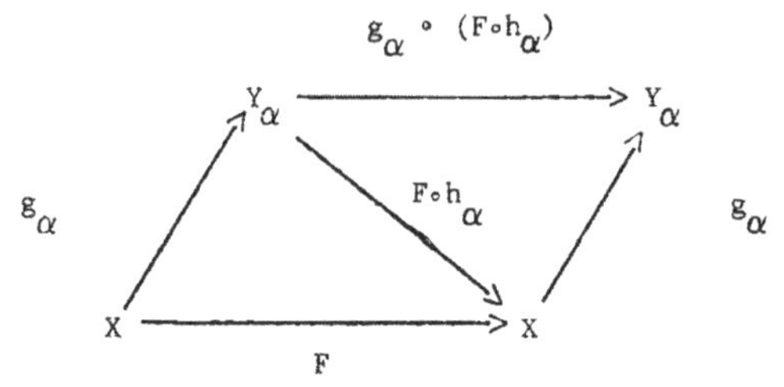

Since F is homotopic to $(F\circ h_\alpha)\circ g_\alpha$ under an acyclic homotopy (compose F with the homotopy from $h_\alpha \circ g_\alpha$ to 1_X), the diagram commutes at the homology level. And since $g_\alpha \circ (F\circ h_\alpha)$ is acyclic and compact, $(g_\alpha)_* \circ (F\circ h_\alpha)_*$ is of finite type. By (5.3), $(g_\alpha)_* \circ (F\circ h_\alpha)_* = F_*$ is of finite type and $\Lambda(F) = \Lambda(g_\alpha \circ (F\circ h_\alpha))$.

Suppose that $\Lambda(F) \neq 0$. Then for each α, there is a point $y_\alpha \in Y_\alpha$ such that $y_\alpha \in g_\alpha(F(h_\alpha(y_\alpha)))$. Let $x_\alpha = h_\alpha(y_\alpha)$ and pick $\bar{x}_\alpha$ in $F(h_\alpha(y_\alpha)) = F(x_\alpha)$ such that $g_\alpha(\bar{x}_\alpha) = y_\alpha$. Then $\{x_\alpha : \alpha \in \text{Cov } X\}$ is a net in X and $\{\bar{x}_\alpha : \alpha \in \text{Cov } X\}$ is a net in $\overline{F(X)}$, a compact set. There is a subnet T of $\{\bar{x}_\alpha\}$ converging to some point $x_0 \in \overline{F(X)}$. We define S to be the corresponding subnet of $\{x_\alpha\}$. Using the fact that $h_\alpha g_\alpha(\bar{x}_\alpha) = x_\alpha$ and the fact that $h_\alpha \circ g_\alpha$ and 1_X are α-near, it can be shown that S also converges to x_0. (Regularity is used here.) Then $S \times T$ is a net in $\Gamma(F)$ converging to (x_0, x_0). Since F is continuous, $\Gamma(F)$ is closed and $(x_0, x_0) \in \Gamma(F)$. Thus x_0 is a fixed point for F.

(6.3) <u>Corollary</u>. A retract of an $M\Lambda$- space is an $M\Lambda$-space.

(6.4) <u>Theorem</u>. A polyhedron P with the Whithead topology is an $M\Lambda$-space.

If $F: P \to P$ is compact and acyclic, then there is a finite subpolyhedron P' of P containing $F(P)$ (see [3]). Let $F': P' \to P'$ be defined by F. Then Eilenberg and Montgomery proved that F' must be a Lefschetz map. Using the following diagram and an argument related to the one above, it can be shown that F is a Lefschetz map.

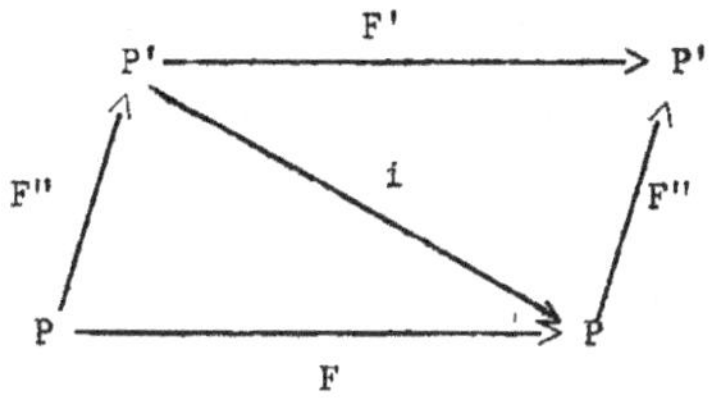

(F'' is defined by F and i is the inclusion.)

(6.5) <u>Corollary</u>. Metric ANR's are Λ-spaces.

For each open cover α of X there is a polyhedron Y_α and maps $g_\alpha\colon X \to Y_\alpha$ and $h_\alpha\colon Y_\alpha \to X$ such that $h_\alpha \circ g_\alpha$ is homotopic to 1_X under an α-homotopy. In particular $h_\alpha \circ g_\alpha$ and 1_X are α-near. (See [3].) Then by (6.2) for Λ-spaces, X is a Λ-space. (Note that (6.2) for Λ-spaces does not require condition (i).)

(6.6) <u>Remark</u>. It has not yet been proved that metric ANR's are $M\Lambda$-spaces. In the proof of (6.5) the spaces Y_α are nerves of covers of X and the maps g_α are not in general acyclic on sets.

A second question involves the relation between Λ and $M\Lambda$-spaces. Clearly, $M\Lambda$-spaces are Λ-spaces. But do there exist Λ-spaces which are not $M\Lambda$-spaces?

The following theorem is proved by a different method and is listed here in order to show an application in §7.

(6.7) <u>Theorem</u>. Every open subset of a Banach space is an $M\Lambda$-space.

7. The following three results are representative of the types of applications of this theory.

(7.1) <u>Schauder Theorem</u>. If X is a convex subset of a Banach space which is either open or closed and $F\colon X \to X$ is compact and acyclic, then F has a fixed point. If F is single-valued, the assumption that X is either open or closed can be omitted. (In the multi-valued case, X is an $M\Lambda$-space by (6.3) and (6.7). In the single-valued case, X is a Λ-space since X is an ANR.)

(7.2) Rothe Theorem. Let B denote a closed ball in a Banach space E and let S denote the boundary of B. If $F: B \to E$ is compact and acyclic and if $F(S) \subseteq B$, then F has a fixed point.

(7.2) is proved by taking a closed ball B' in E such that $B' \supseteq B \cup F(B)$ Then there is a natural retraction $r: B' \to B$ such that $r(B' - B) \subseteq S$. Then $F \circ r: B' \to B'$ is a compact acyclic map on a closed convex subset of a Banach space. Thus $F \circ r$ has a fixed point, which must be a fixed point for F.

The third example illustrates the use of cross-sections of multi-valued maps.

(7.3) Sweeping Theorem. Let B^{n+1} denote the closed unit ball in R^{n+1} and S^n the boundary of B^{n+1}. Let $f: S^n \to S^n$ be a continuous map without fixed points. For each x in S^n join x and $f(x)$ by an acyclic subset $A(x)$ of R^{n+1} in a "continuous" manner. ($A: S^n \to R^{n+1}$ is an u.s.c. acyclic multi-valued map.) Then $B^{n+1} \subseteq A(S^n)$ if n is even.

Suppose that some point x in B^{n+1} is left uncovered by A. Then there is a small closed ball B_1 such that $x \in$ interior B_1 and $B_1 \cap A(S^n) = \emptyset$. Let B_2 be a closed ball in R^{n+1} containing $B^{n+1} \cup A(S^n)$ and let $D = B_2 - B_1$. Maps A', f', $1'_{S^n}$ are defined by $A, f, 1_{S^n}$. Then $f'_* = 1'_*$ since f' and $1'$ are cross-sections of the acyclic map A'. Finally, if $r: D \to S^n$ is the natural retraction $f_* = r_* \circ f'_* = r_* \circ 1'_* = 1_*$. Then $\Lambda(f) = \Lambda(1) = 2$, contradicting the fact that f has no fixed point.

Bibliography

[1] Begle, E. G., "The Vietoris mapping theorem for bicompact spaces", *Am. of Math.*, 51 (1950) 534-543.

[2] Eilenberg, S. and D. Montgomery, "Fixed point theorems for multi-valued transformations", *Am. J. of Math.*, 58 (1946) 214-222.

[3] Hu, S. T., *Theory of Retracts*, Wayne State Univ. Press, 1965.

[4] Jaworowski, J. W. and M. J. Powers, "Λ-spaces and fixed point theorems," *Fund. Math.*, 64 (1969) 157-16?.

[5] Powers, M. J., Thesis, Indiana University, 1968.

[6] ____________, "Multi-valued mappings and Lefschetz fixed point theorems", *Proc. Cambr. Philos. Soc.*, to appear.

EXTREME POINTS OF CONVEX SETS AND SELECTION THEOREMS

by

KONDAGUNTA SUNDARESAN

Selection theorems are useful in characterizing extreme points of convex sets in function spaces. In the present note we illustrate this assertion by characterizing extreme points of the unit cell in Lebesgue-Bochner function spaces. The principal tool in the proof is a selection theorem for upper semi-continuous set valued mappings, Kuratowski and Ryll-Nardzewski [1].

We adhere to the following notation throughout the paper. (X, Σ, μ) is a fixed measure space with X a locally compact Hansdorff space, Σ the σ-ring of Borel sets in X and μ a regular positive measure. For a definition of these terms we refer to Halmos [2]. If E is a locally convex topological vector space a function $f: X \to E$ is said to be measurable if f has the Lusin property i.e. if K is a compact subset of X and ϵ is a positive number then there exists a compact set $C \subset K$ such that $\mu(K \sim C) < \epsilon$ such that $f|C$ is continuous. If E is a Banach space a function $f: X \to E$ is said to be measurable if it is measurable with respect to the strong topology on E. A function $f: X \to E$ is said to be $W(W^*)$ measurable if f is measurable with respect to the weak topology (or weak* topology if appropriate) on E. If E is a Banach space the linear space of measurable functions f such that $x \to \|f(x)\|^p$ $(1 \leq p \leq \infty)$ is μ-summable is denoted by L_E^p. After the usual identification of functions agreeing a.e. it is verified that L_E^p is a Banach space when equipped with the norm $\|f\| = [\int_X \|f(x)\|^p d\mu(x)]^{1/p}$. Likewise L_E^∞ is the Banach space of μ-essentially bounded measurable functions f on X to E with the norm $\|f\| = \operatorname{ess\,Sup}_{x \in X} \|f(x)\|$ we denote the norm in L_E^p $(1 \leq p \leq \infty)$ and the norm in E by the same symbol $\| \;\|$ as there is no occasion for confusion. U_E^p is the unit cell in L_E^p and $U_E(S_E)$ is the unit cell (unit sphere) in E. If C is a convex set $\operatorname{Ext} C$ is the set of extreme points of C. If $K \subset X$ then χ_K is the characterisitc function of K. We proceed to characterize the extreme points in U_E^∞ when E is a separable conjugate Banach space. The proof is rather general and could be adopted to characterize extreme points of U_E^p. See for these and related results Sundaresan [3]. We proceed to the main theorem after stating a lemma and the selection theorem referred to in the introduction.

Lemma 1. If E is a Banach space and $f \in \text{Ext } U_E^\infty$ then $\|f(x)\| = 1$ for x μ a.e.

Proof. Let $f \in \text{Ext } U_E^\infty$. It is easily verified that if $f: X \to E$ is measurable then the function $x \to \|f(x)\|$ is also measurable. Thus if the conclusion is false then there exists a Borel set M in X such that $\mu(M) > 0$ and for all $x \in M$, $\|f(x)\| < 1$. Since μ is a regular measure there exists a compact set $C \subset M$ of positive measure such that $f|C$ is continuous. Hence $f(C)$ is a compact set in the interior of U_E. Thus there exists a vector $V \in U_E$ (choose for V any vector such that $0 < \|V\| < 1\text{-}\underset{x \in C}{\text{Max}} \|f(x)\|$) such that $\|f(x) \pm V\| \leq 1$ for all $x \in C$. Let now g_i, $i = 1, 2$, be the functions $(f \pm V)\chi_C + f\chi_{C'}$. It is verified that $\|g_i\| = 1$, $f = \frac{g_1 + g_2}{2}$ and $g_1 \neq g_2$. Thus $f \notin \text{Ext } U_E^\infty$ completing the proof of the assertion.

Concerning set valued mappings we recall a definition and a useful selection theorem. Let X,Y be two topological spaces and 2^Y be the set of all closed subsets of Y. A function $F: X \to 2^Y$ is called upper semi-continuous (u.s.c.) if the set $\{x | F(x) \subset G\}$ is open in X for all open sets G in Y. If X,Y are two topological spaces a function $f: X \to Y$ is called Borel measurable if $f^{-1}(G) \in \Sigma$ for all open sets G in Y where Σ is the σ-ring generated by open sets in X. We state a selection theorem due to Kuratowski and Ryll-Nardzewski [1] in a form suitable for our purpose here.

Theorem [Kuratowski and Ryll-Nardzewski]. Let X be a metric space and (Y,d) be a separable metric space which is d-complete. If $F: X \to 2^Y$ is a u.s.c. map with $F(x) \neq \emptyset$ for all $x \in X$ then there exists a Borel measurable function $f: X \to Y$ such that $f(x) \in F(x)$ for each $x \in X$.

We need also a property of Banach spaces which we state here. If E is a conjugate separable Banach space then W*- topology relativised to the unit cell in E is metrizable. We can further assume that such a metric d on U_E could be defined to satisfy $d(p,q) \leq \|p-q\|$ for $p,q \in U_E$. For if $E = B^*$ then B is also a separable Banach space. Thus there exists a countable dense subset $\{x_n\}_{n\geq 1}$ of the unit cell U_B with respect to the norm topology relativised to U_B. For p,q in U_E define $d(p,q) = \sum_{n\geq 1} \frac{1}{2^n} |p(x_n) - q(x_n)|$. Then d has the required properties. See in this connection Theorem 1, p. 426, Dunford and Schwartz [4].

Theorem 1. If E is a conjugate separable Banach space then a function $f \in L_E^\infty$ is in Ext U_E^∞ if and only if $f(x) \in \text{Ext } U_E$ for x μ a.e.

<u>Proof.</u> To prove the "only if" part let us start noting that since U_E is a compact convex metrizable subset of E in the W*-topology the set Ext U_E is a W*-G_δ subset of U_E. See proposition 1.3, Phelps [5]. Hence Ext U_E is a Borel set in the norm topology of E. Further from Lemma 1 it follows that $\|f(x)\| = 1$ μ a.e. Thus if $f(x) \notin \text{Ext } U_E$ for μ a.e. then there exists a Borel set $M \subset X$ with $\mu(M) > 0$ such that for all $x \in M$, $f(x) \notin \text{Ext } U_E$. Since μ is a regular measure there exists a compact set $C \subset M$, $0 < \mu(C) < \infty$ such that $f|C$ is continuous. Hence $f(C)$ is a compact in S_E. We note that for $x \in C$ there exist $p_x, q_x \in S_E$ such that $p_x \neq q_x$ and $f(x) = \frac{p_x + q_x}{2}$. For positive δ let C_δ be the subset of C such that there exist $p_x, q_x \in S_E$ with $f(x) = \frac{p_x + q_x}{2}$, $d(p_x, q_x) \geq \delta$ where d is the metric defined in the paragraph preceeding the statement of the theorem. We proceed to verify C_δ is a Borel set

in X, in fact a closed subset of the compact set C. For let $\{t_n | n \in D\}$ be a net in C_δ such that $t_n \to t$ for some $t \in C$. Let $f(t_n) = \frac{p_n + q_n}{2}$, for some $p_n, q_n \in S_E$ such that $d(p_n, q_n) \geq \delta$. Since (U_E, d) is a compact metric space there exist convergent subnets $\{p_{ni}\}, \{q_{ni}\}$ in $\{p_n\}, \{q_n\}$ respectively. From the continuity of $f|C$ it follows that $f(t) = \frac{p+q}{2}$ if $p_{n_i} \to p$ and $q_{n_i} \to q$. Further $d(p_{n_i}, q_{n_i}) \geq \delta$ implies $d(p,q) \geq \delta$. Thus $t \in C_\delta$ and C_δ is a closed subset of C. By considering the sequence of disjoint Borel sets $\{C_{\frac{1}{m}} \sim C_{\frac{1}{m+1}}\}$ for integers $m \geq 1$ and noting that $0 < \mu(C) < \infty$ it follows that there exists a compact set $C_o = C_{2k} \subset C$ and two functions $f_i: C_o \to S_E$, $i = 1, 2$ such that for all $x \in C_o, f(x) = \frac{f_1(x) + f_2(x)}{2}$ and $\|f_1(x) - f_2(x)\| \geq d(f_1(x), f_2(x)) \geq 2k$. Thus there exists a function $F: f(C_o) \to 2^{U_E}$ $F(\xi)$ being the nonempty W^*-closed set of points α in S_E satisfying the condition for some $\beta \in S_E$, $\xi = \frac{\alpha + \beta}{2}$ and $\|\alpha - \beta\| \geq d(\alpha, \beta) \geq 2k$. With the norm topology on E relativised to $f(X_o)$ and W*-topology relativised to U_E we proceed to verify that F is a u.s.c. map. Let G be an open set in (U_E, d) and let $G_1 = \{\xi | F(\xi) \subset G\}$.

Suppose that $t \in f(C_o)$ and that there is no neighborhood N of t such that for all $\eta \in N$, $F(\eta) \subset G$. Thus there exists a sequence $\{t_n\}$ in $f(C_o)$, $t_n \to t, F(t_n) \not\subset G$ for all $n \geq 1$ which in turn implies the existence of a sequence $\{t_n^1\}$ in S_E with $t_n^1 \in F(t_n) \sim G$. Considering a sequence $\{t_n^2\}$ in U_E such that $t_n = \frac{t_n^1 + t_n^2}{2}$ $d(t_n^1, t_n^2) \geq 2k$ by standard compactness arguments it follows that there exists a subsequence $\{t^1_{n_i}\}$ in $\{t^1_n\}$ converging to some point t^1 in the space (U_E, d) such that for some point $t^2 \in U_E$, $t = \frac{t^1 + t^2}{2}$ and $d(t^1, t^2) \geq 2k$. Thus $t^1 \in F(t) \subset G$. Since G is a neighborhood of t^1 there exist $t_{n_i} \in G$ leading to a contradiction. Thus F is a u.s.c. map. It is verified that $f(C_o)$, (U_E, d) satisfy the conditions on X and Y in the selection theorem stated earlier in the paper. Hence there exists a Borel measurable function $h: f(C_o) \to (U_E, d)$ such that $h(\xi) \in F(\xi)$ for all $\xi \in f(C_o)$. Let $g^1: C_o \to U_E$ be the function defined by $g^1 = h \circ f$. Since E is a separable Banach space by the theorem 3.5.(2) on p. 74 in [5] it follows that g^1 is a measurable function. Further from the definition of F and the choice of h it is inferred that there exist $g^2(x) \in S_E$ such that $f(x) = \frac{g^1(x) + g^2(x)}{2}$, $\|g^1(x) - g^2(x)\| \geq d(g^1(x), g^2(x)) \geq 2k$. Since f, g^1 are measurable, g^2 is a measurable function and $f|C_o = \frac{g^1 + g^2}{2}$. Now defining the functions f^i, $i = 1.2$ by the equations $f^i(x) = g^i(x)$ if $x \in C_o$ and $f^i(x) = f(x)$ if $x \in X \sim C_o$ it is verified that $f^i \in U^\infty_E$, $f = \frac{f^1 + f^2}{2}$ and $f^1 \neq f^2$. Thus $f \in \text{Ext}\, U^\infty_E$ contradicting the choice of f.

The "if" part is evident for if $f \in U^\infty_E \sim \text{Ext}\, U^\infty_E$ then there exist $f^1, f^2 \in U^\infty_E$, $f^1 \neq f^2$ such that $f(x) = \frac{f^1(x) + f^2(x)}{2}$ for x μ a.e. If $M = \{x | f^1(x) \neq f^2(x)\}$ then M is a Borel set in X and $\mu(M) > 0$ and for all $x \in M$, $f(x) \notin \text{Ext}\, U_E$.

Thus the proof of the theorem is completed.

The characterization of $\text{Ext}\, U^\infty_E$ in the Theorem 1 is useful in the theory of operators. For this we refer to [3].

BIBLIOGRAPHY

1. K. Kuratowski and C. Ryll-Nardzewski, A General Theorem on Selectors, Bull. Acad. Polon. Sci. Ser. Sci. Math. Astronom. Phys., 13 (1965), 397-403.

2. P. R. Halmos, "Measure Theory", D. Van Nostrand, New York, 1950.

3. K. Sundaresan, Extreme Points of Convex Sets in Lebesgue-Bochner Function Spaces II Report 69-13, Department of Mathematics, Carnegie-Mellon University (1969).

4. N. Dunford and J. T. Schwartz, Linear Operators Part I, Interscience Publishers, New York, 1958.

5. E. Hille and R. S. Phillips, "Functional Anaylysis and Semi-groups" Amer. Math. Soc. Coll. Publ. XXXI, 1957.

SET-VALUED MAPPINGS ON PARTIALLY ORDERED SPACES

by

L. E. Ward, Jr.

1. Introduction. We are concerned here with a pair (X, Γ) where X is a topological space and $\Gamma : X \to X$ is a set-valued mapping satisfying

(i) $x \in \Gamma x$ for all $x \in X$,

(ii) $\Gamma(\Gamma x) \subset \Gamma x$ for all $x \in X$,

(iii) Γ is univalent.

If we write $x \leq y$ for $x \in \Gamma y$ then (i), (ii), and (iii) translate into the reflexive, transitive and asymmetric laws, respectively, for the binary relation $\leq$. Accordingly we call Γ a partial order, and if Γ has a closed graph then (X, Γ) is a partially ordered space.

The theory of partially ordered spaces has undergone extensive development since about 1947, beginning with the early work of Choquet [2] and Nachbin [9, 10, 11, 12]. Its literature is widely scattered and diverse and it is not possible to give a comprehensive survey here. An incomplete (but representative) bibliography is appended to this article. Our present purpose is to give a summary of four "chapters" of the theory which have been of especial interest to the author. In each case we will see that the theorems developed have applications of interest which are external to the theory itself.

A few notational conventions should be summarized at this point. As noted above, the expressions "$x \leq y$" and "$x \in \Gamma y$" are synonyms; we also write $y \in x\Gamma$ to mean $x \in \Gamma y$, and if $A \subset X$ then

$$\Gamma A = \cup \{\Gamma x : x \in A\},$$

$$A\Gamma = \cup \{x\Gamma : x \in A\}.$$

A set A is increasing (decreasing) if $A = A\Gamma$ ($A = \Gamma A$). Decreasing sets are also called _ideals_, and a set Γx (where $x \in X$) is a _principal ideal_. An element x of X is _maximal_ (minimal) if x has no proper Γ - successor (-predecessor), and the set of maximal (minimal) elements of X is denoted Max X (Min X). A _zero of_ (X, Γ) is an element $0 \in X$ such that $0\Gamma = X$.

The following elementary propositions follow from these definitions.

Proposition 1.1. [24]. _A partially ordered space is a Hausdorff space._

Proposition 1.2. [20]. _If_ (X, Γ) _is a partially ordered space and_ $x \in X$ _then_ Γx _and_ $x\Gamma$ _are closed sets._

Proposition 1.3. [20]. _Each maximal chain of a partially ordered space is a closed set._

Proposition 1.4. [20]. _If_ X _is a compact partially ordered space then_ Max X _and_ Min X _are non-empty._

Proposition 1.5. [2, 27]. _If_ (X, Γ) _is a compact partially ordered space and_ A _is a closed subset of_ X _then_ ΓA _is a closed set._

Proposition 1.6. _If_ (X, Γ) _is a compact partially ordered space and_ x _and_ y _are members of_ X _such that_ $(x, y) \in X \times X - \Gamma$, _then there exists an open set_ U _such that_ $x \in U = U\Gamma$ _and_ $y \in X - \overline{U}$.

2. Some descendents of the Scherrer fixed point theorem. Let X be a connected topological space. An element $p \in X$ is a _cutpoint_ of X provided $X - \{p\}$ is not connected. If p is a cutpoint of X then p _separates_ the elements a and b provided a and b lie in distinct components of $X - \{p\}$. If we fix $e \in X$ and define $x \leq y$ if and only if $x = e$ or $x = y$ or x separates e and y, then the following proposition is easily verified.

Proposition 2.1. [24]. _The relation_ $\leq$ _is a partial order._

This partial order is called the _cutpoint order_ on the connected space X and hereafter we denote it Γ_e.

Proposition 2.2. [25]. *If* x *and* y *are elements of* X *then* $\Gamma_e x \cap \Gamma_e y$ *is a nonempty chain.*

Proposition 2.3. Max X *consists of all non-cutpoints other than* e.

Unfortunately Γ_e does not always have a closed graph; however we do have the following result. (In this paper a *continuum* is a compact connected Hausdorff space. A continuum is *semi-locally connected* provided each point is contained in arbitrarily small open sets whose complements have only finitely many components.)

Theorem 1. [30]. *If* X *is a semi-locally connected continuum and* $e \in X$ *then* Γ_e *has a closed graph. Moreover, if* $x \in X$ *then* $x\Gamma_e - \{x\}$ *is an open set.*

A *tree* is a continuum with the property that each two distinct points are separated by some third point. These objects have been extensively studied (see, for example, [31]). The metric trees (or *dendrites*) are simply the Peano continua which contain no simple closed curves. Using Theorem 1 we obtain another characterization.

Theorem 2. [25]. *Let* X *be a compact Hausdorff space. A necessary and sufficient condition that* X *be a tree is that* X *admit a partial order* Γ *satisfying*

(i) Γ *has a closed graph,*

(ii) Γ *is dense,* i.e., *if* x *and* z *are members of* X *with* $x < z$ *then there exists* $y \in X$ *such that* $x < y < z$,

(iii) *if* x *and* y *are elements of* X *then* $\Gamma x \cap \Gamma y$ *is a nonempty chain,*

(iv) *if* $x \in X$ *then* $x\Gamma - \{x\}$ *is an open set.*

The partial order Γ of Theorem 2 is, of course, the cutpoint order Γ_e for some $e \in X$.

Theorem 3. *If* X *is a tree and* $e \in X$ *then, relative to* Γ_e, *each subcontinuum of* X *has a zero.*

Theorems 2 and 3 can be employed to prove that a tree enjoys two very general fixed point properties. The first result of this type was established in 1926 by W. Scherrer [14] who showed that a dendrite has the fixed point property for homeomorphisms. A comprehensive survey of the many progeny of Scherrer's theorem can be found in the memoir of Vander Walt [18].

Suppose X and Y are spaces and that $F: X \to Y$ is a set-valued mapping. We say that F is *upper semi-continuous* provided F(x) is a closed set for each $x \in X$ and, whenever V is an open set with $F(x) \subset V$, there exists an open set U with $x \in U$ such that $F(t) \subset V$ for each $t \in U$. Further, F is *continuous* if it is upper semi-continuous and if, whenever $x \in X$ and V is an open set which meets F(x), there exists an open set U with $x \in U$ such that V meets F(t) for each $t \in U$.

The next theorem was first proved by Wallace [19], using other methods than the order theoretic techniques developed here.

Theorem 4. *A tree has the fixed point property for the upper semi-continuous continuum-valued mappings.*

By essentially the same techniques we can also obtain a result of Plunkett [13].

Theorem 5. *A tree has the fixed point property for continuous set-valued mappings.*

These results are by no means the most general which can be obtained by order-theoretic arguments. A continuum X is *unicoherent* if, for each two subcontinua A and B such that $X = A \cup B$, it follows that $A \cap B$ is connected. A continuum is *hereditarily unicoherent* if each of its subcontinua is unicoherent. If a metrizable continuum is both hereditarily unicoherent and arcwise connected it is termed a *dendroid.* These objects admit a partial order very similar to that of a tree [26] which permits a proof of the following results.

Theorem 6. *A dendroid has the fixed point property for continuous set-valued mappings.*

Theorem 7. *Among the arcwise connected metrizable continua, the property of being a dendroid is equivalent to having the fixed point property for upper semi-continuous continuum-valued mappings.*

The following problem was first posed in 1961 [26].

Problem. *Among the arcwise connected metrizable continua, is the property of being a dendroid equivalent to having the fixed point property for continuous set-valued mappings?*

I conjecture that the answer is yes.

3. The non-cutpoint existence theorem. One naturally expects the cutpoint order to facilitate the study of spaces which are rich in cutpoints, and this expectation has been born out by Theorems 2, 3, 4, and 5. The cutpoint order can also be employed to give a new proof of the classical non-cutpoint existence theorem. To this end we define a *weak partially ordered space* to be a pair (X, Γ) where X is a topological space, Γ is a partial order on X, and for each $x \in X - \text{Max } X$ there exists a closed set $K(x) \subset x\Gamma$ such that $K(x) - \{x\}$ is a non-empty and increasing. It is easy to see that a partially ordered space is a weak partially ordered space but not conversely. The following results were established in [28].

Proposition 3.1. *If* X *is a compact weak partially ordered space then* Max X *is not empty. Further, if* X *is not a chain and if* Γx *is a chain for each* $x \in X$, *then* Max X *contains at least two elements.*

Proposition 3.2. *If* X *is a connected Hausdorff space and* $e \in X$ *then* (X, Γ_e) *is a weak partially ordered space.*

Theorem 8. _A non-degenerate continuum has at least two non-cutpoints._

Proof. Let X be a non-degenerate continuum. The theorem is obvious if X is cutpoint free so we may assume that X contains a cutpoint e. By Proposition 3.2 (X, Γ_e) is a weak partially ordered space. By a simple argument involving Proposition 2.2, (X, Γ_e) satisfies the hypotheses of Proposition 3.1 and therefore Max X contains at least two maximal elements. The theorem now follows from Proposition 2.3.

4. Theorems on arcs. Another classical problem of topology is to determine when a space is arcwise connected. Here it will be convenient to adopt the terminology of A. D. Wallace [22] and call a subset A of a space an _arc_ if A is a continuum with exactly two non-cutpoints. If A is also separable then it is a _real arc_. An arc which is contained in a partially ordered space in such a way that its natural order coincides with the partial order of the space is called an _order arc_.

A few years ago R. J. Koch [6] proved a remarkable theorem on the existence of arcs. He showed that a compact partially ordered space is arcwise connected, given only mild hypotheses on the principal ideals. In this section we sketch a different proof (see also [27]) and we also give some indication how far reaching Koch's theorem is.

A subset S of a partially ordered space is said to _have no local minima_ provided for each $x \in S$ and each neighborhood U of x there exists $y \in U \cap S$ such that $y < x$.

Proposition 4.1. _Let_ (X, Γ) _be a compact partially ordered space and let_ W _be an open subset of_ X _which has no local minima. Then_ Γ _contains a partial order_ Γ_0 _relative to which_ (X, Γ_0) _is a partially ordered space,_ W _has no local_ Γ_0_-minima, and_ Γ_0 _is minimal relative to these properties._

Proposition 4.2. *Let* X *be a compact space,* W *an open subset of* X, *and* Γ *a partial order relative to which* (X, Γ) *is a partially ordered space,* W *has no local minima, and* Γ *is minimal relative to these properties. Then every maximal chain of* (X, Γ) *is connected.*

Theorem 9. *Let* X *be a compact partially ordered space and let* W *be an open subset of* X *which has no local minima. If* $x \in W$ *then* x *is the supremum of an order arc which meets* $X - W$.

Proof. By Proposition 4.1 we may assume the partial order is minimal. If $x \in W$ we let D be a maximal chain which contains x. By Proposition 4.2 D is an order arc, and since W has no local minima it follows that $D \cap \Gamma x$ is non-degenerate and hence is an order arc whose supremum is x. Finally, the least element of $D \cap \Gamma x$ cannot lie in W.

Corollary 9.1. *If* X *is a compact partially ordered space with zero and if* Γx *is connected for each* $x \in X$, *then* X *is arcwise connected.*

Proof. Let $W = X - \{0\}$ and apply Theorem 9.

Perhaps the most famous theorem on arcwise connectedness is the classical result of R. L. Moore that a Peano continuum is arcwise connected. In [27] it was conjectured that Koch's arc theorem implies this result and in [5] Virginia Walsh Knight proved that this conjecture is correct.

Theorem 10. *If* X *is a Peano continuum and if* $0 \in X$ *then there is a partial order* Γ *such that* (X, Γ) *is a partially ordered space,* 0 *is a zero for* (X, Γ), *and* Γx *is connected, for each* $x \in X$.

5. Some theorems on acyclicity. A *directed space* is a partially ordered space in which the family of principal ideals has the finite intersection property. Thus a tree with the cutpoint order is a directed space, as is any space satisfying the hypotheses of Corollary 9.1.

Corollary 9.1 can be regarded as an acyclicity theorem in the following sense. The assumption that the principal ideals are acyclic in dimension 0 (i.e., they are connected) implies that the space is acyclic in dimension 0 in a strong sense (arcwise connected). It is natural to seek generalizations to higher dimensions, and this final section is devoted to results of that type.

A space is said to be _acyclic in dimension_ n provided it has the SWAK n-th cohomology group of a point [4]. It is _acyclic_ if it is acyclic in all dimensions $0, 1, 2, \ldots$. A partially ordered space is _pointwise acyclic_ if each of its principal ideals is acyclic. Our first result is essentially due to Wallace [21].

Theorem 11. _A compact, pointwise acyclic directed space is acyclic._

Corollary 11.1. _If_ S^n _is partially ordered as a directed space and if each principal ideal of_ S^n _is acyclic in dimensions_ $0, 1, \ldots, n-1$, _then_ S^n _has a unit._

Theorem 12. _Let_ X _be a compact directed space and suppose there exists exactly one element_ $x_1 \in X$ _such that_ Γx_1 _is not acyclic. Then_ $x_1 \in \text{Max } M$ _and_ $H^n(X) \cong H^n(\Gamma x_1)$ _for all_ $n = 0, 1, 2, \ldots$.

Theorem 13. _Let_ X _be a compact directed space which contains exactly_ n _elements_ $x_1, \ldots, x_n$ _such that_ $\Gamma x_1, \ldots, \Gamma x_n$ _are not acyclic. Then these elements can be labeled so that for some_ m, $1 \leq m \leq n$, _the elements_ $x_1, \ldots, x_m$ _are maximal and if_ $m < i \leq n$ _then_ $x_i < x_j$ _for some_ $j \leq m$. _Further, if_ $x_i < x < x_j$ _then_ x _is one of the elements_ $x_1, \ldots, x_m$. Finally $H^p(X)$ _contains each of the groups_ $H^p(\Gamma x_i)$ _as a direct summand, for all_ $p > 0$ _and_ $i \leq m$.

Problem. *Let* X *be a compact directed space and suppose that the set of elements which generate non-acyclic principal ideals is closed and totally disconnected. Does the cohomology of those ideals determine the cohomology of* X?

If F is a closed ideal of a partially ordered space then X/F (the quotient space which results from identifying F with a point) becomes a new partially ordered space in a natural way. This quotient has interesting implications for the study of acyclicity in partially ordered spaces.

Theorem 14. *Let* X *be a compact directed space with only finitely many non-acyclic principle ideals, all of which lie in the closed ideal* F. *Then* X/F *is an acyclic directed space.*

Corollary. *If the* n-*cell* B^n *is partially ordered as a compact directed space in such a way that the bounding* (n - 1) - *sphere is an ideal, then either there exists an element* $x \in B^n - S^{n-1}$ *such that* Γx *is not acyclic or* S^{n-1} *contains infinitely many non-acyclic principle ideals.*

REFERENCES

[1] J. H. Carruth, A note on partially ordered compacta, Pacific J. Math. 24 (1968) pp. 229-231.

[2] G. Choquet, Convergences, Annales Grenoble, Sec. des Sci, Math. et Phys. 23 (1947) pp. 58-112.

[3] S. P. Franklin and A. D. Wallace, The least element map, Colloq. Math. 15 (1966) pp. 217-221.

[4] S. T. Hu, Cohomology Theory, Chicago, 1968.

[5] Virginia Walsh Knight, A continuous partial order for Peano continua, Pacific J. Math., 30 (1969) pp. 141-153.

[6] R. J. Koch, Arcs in partially ordered spaces, Pacific J. Math. 9 (1959), pp. 723-728.

[7] I. S. Krule, Structs on the 1-sphere, Duke Math. J. 24 (1957), pp. 405-414.

[8] S.-Y. T. Lin, A characterization of the cutpoint order on a tree, Trans. Amer. Math. Soc. 124 (1966), pp. 552-557.

[9] L. Nachbin, Sur les espaces topologique ordonnés, C.R. Acad. Sci. Paris 226 (1948), pp. 381-382.

[10] ________, Sur les espaces uniformisables ordonnés, Ibidem 226 (1948), p. 547.

[11] ________, Sur les espaces uniformes ordonnés, ibidem 226 (1948), pp. 774-775.

[12] ________, Topology and Order, Princeton, 1964.

[13] R. L. Plunkett, A fixed point theorem for continuous multi-valued transformations, Proc. Amer. Math. Soc. 7 (1956), pp. 160-163.

[14] W. Scherrer, Uber ungeschlossene stetige kurven, Math. Zeit. 24 (1926), pp. 125-130.

[15] R. E. Smithson, A note on acyclic continua, Colloq. Math. 19 (1968), pp. 67-71.

[16] E. D. Tymchatyn, The 2-cell as a partially ordered space, Pacific J. Math., 30 (1969), pp. 825-836.

[17] ____________, and L. E. Ward, Jr., On three problems of Franklin and Wallace concerning partially ordered spaces, Coll. Math., 20 (1969) pp. 229-236.

[18] T. Van der Walt, Fixed and Almost Fixed Points, Amsterdam, 1963.

[19] A. D. Wallace, A fixed point theorem for trees, Bull. Amer. Math. Soc. 47 (1941), pp. 757-760.

[20] ____________, A fixed point theorem, ibidem 51 (1945), pp. 413-416.

[21] ____________, A theorem on acyclicity, ibidem 67 (1961), pp. 123-124.

[22] ____________, Relations on topological spaces, Proc. Symp. on Gen. Topology and its Relations to Modern Analysis and Algebra, Prague, 1961, pp. 356-360.

[23] A. J. Ward, A theorem of fixed point type for non-compact locally connected spaces, Colloq. Math. 17 (1967), pp. 289-296.

[24] L. E. Ward, Jr., Partially ordered topological spaces, Proc. Amer. Math. Soc. 5 (1954), pp. 144-161.

[25] ____________, A note on dendrites and trees, ibidem 5 (1954), pp. 992-994.

[26] ____________, Characterization of the fixed point property for a class of set-valued mappings, Fund. Math. 50 (1961), pp. 159-164.

[27] ____________, Concerning Koch's theorem on the existence of arcs, Pacific J. Math. 15 (1965), pp. 347-355.

[28] ____________, On the non-cutpoint existence theorem, Can. Math. Bull. 11 (1968), pp. 213-216.

[29] ____________, Compact directed spaces, Trans. Amer. Math. Soc., to appear.

[30] ____________, A general fixed point theorem, Colloq. Math. 15 (1966), pp. 243-251.

[31] G. T. Whyburn, Analytic Topology, New York, 1942.

FACTORS OF SUBCLASSES OF $2^{(I^m)}$

by

Raymond Y. T. Wong

If X is a metric space, the hyperspace of X, denoted 2^X, is the space of all non-void closed subsets of X with the usual Hausdorff metric. The n-fold ($n \geq 1$) symmetric product (Borsuk-Ulam [1]) of X, denoted $X(n)$, is the subspace of 2^X consisting of all elements with $\leq n$ points. Let I denote the closed unit interval, I^n the n-cube and I^∞ the Hilbert cube. Let J^∞ denote another copy of the Hilbert cube with $J = [-1, 1]$ and let R be the equivalence relation on J^∞ defined by identifying each $x = (x_1, x_2, \ldots)$ with $-x = (-x_1, -x_2, \ldots)$. Let $S(X)$ denote the subspace of 2^X consisting of all continua. In [3] R. Schori shows that for $n \geq 1$ and $\alpha = \infty, 1, 2, \ldots$, $I^\alpha(n)$ contains I^α as a factor; that is, $I^\alpha(n)$ is homeomorphic to $Y \times I^\alpha$ for some space Y.

Theorem I. Let m,n be positive integers. If $X = I^m(n)$, $2^{(I^m)}$ or $S(I^m)$, then X contains I^m as a factor.

Thus we generalize Schori's Theorem in the case m is finite. In the case when $m = \infty$, we have

Theorem II. If $X = I^\infty(n)$, $S(I^\infty)$ or 2^{I^∞}, then for any positive integer k, X contains I^k as a factor.

Question. If $X = S(I^\infty)$ or 2^{I^∞}, must X contain I^∞ as a factor?

Question. Is $J^\infty(2)$ homeomorphic to J^∞?

It is known that $J^\infty(2)$ is homeomorphic to $J^\infty/_R \times J^\infty$ [3] and $J^\infty/_R$ is not homeomorphic to J^∞.

REFERENCES

[1] K. Borsuk and S. Ulam, On symmetric products of topological spaces, Bull. A.M.S., 37 (1931), 875-882.

[2] V. L. Klee, Jr., Some topological properties of convex sets, Trans. Amer. Soc. 78 (1955) 30-45.

[3] R. Schori, Hyperspaces and symmetric products of topological spaces. Fund. Math. LXIII (1968) 77-88.

CONTINUOUS SELECTIONS ON LOCALLY COMPACT SEPARABLE METRIC SPACES

by

Gail S. Young

In this discussion, I will review the contents of two papers which will appear elsewhere, both of which are concerned with the restrictions on a space which are imposed by the existence of continuous selections on spaces of closed subsets of the space. Hoping that it may be helpful to the general reader of this volume, I will include in the written version some general exposition inappropriate for the original audience of specialists.

Let X be a topological space. We denote by 2^X the collection of all closed subsets of X with the Vietoris topology. Following E. Michael [7] we define this as follows: Let U be an open set in X. By $B_1(U)$ we mean the collection of all closed subsets of X that lie in U. By $B_2(U)$ we mean the collection of all closed sets of X that intersect U. We take the collection B of all sets $B_1(U)$, $B_2(U)$ to be a subbasis for 2^X. It is easy to see that if U, V are open sets in X, then (a) $B_1(U) \cap B_1(V) = B_1(U \cap V)$, and (b) $B_2(U) \cap B_2(V) = B_2(U \cap V)$. However, if we consider an element C of $B_1(U) \cap B_2(V)$ there is "in general" no open set W of X such that either $B_1(W)$ or $B_2(W)$ contains C and lies in $B_1(U) \cap B_2(V)$. Thus B is not in general a basis.

If X is a compact metric space, the Hausdorff metric for the set of all closed subsets of X is a metric that induces the Vietoris topology.

The notion of limit point in 2^X given by the Vietoris topology will be clarified by the following example: Let $I_0, I_1, I_2, I_3, \ldots$ be the closed intervals in the plane of length 1, and perpendicular from above to the x-axis at the respective points (0, 0), (1, 0), (1/2, 0), (1/3, 0), ... Let $X = \bigcup_{n=0}^{\infty} I_n$.

Let A be the point-set in 2^X whose points are the sets $I_1, I_2, I_3, \ldots$.

We ask if the interval I in I_0 from $(0, 0)$ to $(0, 1/2)$ is a limit point of A. If U is any open set in X that contains I, then U meets all but a finite number of the intervals I_n, so that $B_2(U)$ contains almost all points I_n. But if U is the set of all points in X of ordinate $< 2/3$, then $B_1(U)$ contains I, but no point of A. Hence I is not a limit point of A.

Let J_n, $n = 1, 2, 3, \ldots$, be the closed subinterval of I_n of height $1/2$ and with one end of the x-axis. Let Y denote the subset of 2^2 whose points are the sets J_n. Is I_0 a limit point of Y? If U is any open set that contains I_0, then $B_1(U)$ contains all but a finite number of points J_n. However, if U is the set of all points in X of ordinate $> 2/3$, $B_2(U)$ contains I_0, but no point of Y. The only limit point of Y is the interval from $(0, 0)$ to $(0, 1/2)$.

From the Axiom of Choice, we know that there is a function $f\colon 2^X \to X$ which assigns to each closed set of X a point of itself. Such a function is a _selection._ We are concerned with conditions under which such a selection function is _continuous_ in the Vietoris topology. That such continuous selections exist is apparent; for example, let X be the unit interval, and assign to each closed set of X its leftmost point (or rightmost point).

Or let X be any infinite set with the discrete topology. The space 2^X is not discrete; any infinite set is a limit point of the set of its subsets. Well order X and define $\beta\colon 2^X \to X$ by mapping each subset of X into its first element in the well-ordering.

However, as Engelking, Heath and Michael [2] have shown, there is no continuous selection on 2^R, the space of closed subsets of the reals. An argument to show this is the following: Suppose $f\colon 2^R \to R$ is a continuous selection. Let Z denote the integers, and let $f(Z) = n$. There is a deformation $\varphi\colon Z \times I \to R$ defined by $\varphi(m, t) = m$, for $m \neq n$; $\varphi(n, t) = (1 - t)n + t(n - 1)$. An easy argument of continuity shows that for each t, $0 \leq t \leq 1$, $f[(Z - n) \cup \varphi(n, t)] = \varphi(n, t)$. Thus $f(Z - n) = n - 1$. But there is a similar deformation of Z onto

$Z - n$ which maps n onto $n \neq 1$, and the same argument show that $f(Z - n) = n + 1$, a contradiction.

If we had restricted ourselves to the subspaces of 2^R composed of all compact sets, or perhaps of all finite sets, the greatest-lower-bound selection would be well defined and continuous. This fact provides a motivation for our next definition: If X is a topological space, by $F_2(X)$ we mean the subspace of 2^X composed of all sets of one or two points. In a Hausdorff space, $F_2(X)$ is a closed subset of 2^X.

The first topic on which I wish to report is joint work with K. Kuratowski and S. B. Nadler [5] to appear in Fundamenta Mathematica. In this paper, we begin by supposing that we have a metric continuum M--a compact, connected metric space --and that there is a continuous selection f on $F_2(M)$, and ask what restrictions this imposes on M. We were somewhat surprised to find that the only space which satisfies this condition is the unit interval.

Theorem 1. *If* M *is a metric continuum, and* $f\colon 2^M \to M$ *is a continuous selection, then* M *is an arc.*

I will not attempt to give the details of the argument here. Actually, we give two proofs, one depending on a theorem of Borsuk and Mazurkiewicz [4, p. 187] and the other on R. L. Moor's characterization of an interval as a metric continuum with just two non-cutpoints [3, p. 53]. I will give something of the flavor of the proof.

An essential tool in the first proof is the following.

Lemma: *Let* X *be composed of an arc* ab *and of countably many points,* p_1, p_2, p_3, ... *lying outside this arc and converging to a non-end point* c *of* ab. *Then there is no continuous selection on* $F_2(X)$.

Proof. Suppose f is a continuous selection on $F_2(X)$. Suppose, without loss, that $f(\{a,b\}) = a$. The sort of deformation and continuity argument given in the above example, used this time by "sliding" b to x, shows that if x is in ab - a, then f({a, x}) = a. In particular, f({a, c}) = a. A similar argument shows that if x is in ab - b, then f({x, b}) = x; in particular, f({c, b}) = c.

Now for n fixed, we have two choices for $f(\{p_n, a\})$. If (1) the selection is a, we conclude by sliding a to b that $f(\{p_n, b\}) = b$. The other choice is (2) the selection is p_n.

The same one of the two choices is made for infinitely many pairs $\{p_n, a\}$. If it is the first, we conclude from $f(\{p_n, b\}) = b$ and continuity that f({c, b}) = b, which is false. If it is the second, we conclude from $f(\{p_n, a\}) = p_n$ and continuity that f({c, a}) = c, which also is false, and completes the proof.

By considering all pairs of diametrically opposite points and using an argument like that which shows that $\sqrt{z}$ is not single-valued, one can show easily that there is no continuous selection on $F_2(S^1)$, S^1 being the 1-sphere. If we knew that the continuum M were locally connected, we would then know that it contained no simple closed curve, and from the lemma that it contained no triod. Hence it would be an arc. This is as close as I will come to proving Theorem 1.

The first paper I know of concerning continuous selections is that of Michael [7] in 1951. Reexamining his paper after proving Theorem 1, we realized that he came quite close to Theorem 1, and in fact could easily have stated it as a corollary to some of his results. He also proves results which let one say at once that the only continuous selections on $J_2(I^1)$ are (1) assign to each pair the lesser element, or (2) assign to each pair the greater. Indeed he shows that these g.l.b. and l.u.b. selections are the only ones on 2^{I^1}. Michael's paper is fundamental to an understanding of continuous selections, and contains problems still unsettled.

From Theorem 1, we went on to consider what could happen if the space M had more than one component. This situation becomes complicated by the fact that there is then no "natural" selection. Consider, for example, $M = [0, 1] \cup [2, 3]$. We have the l.u.b. and g.l.b. selections. But there are others. For example, define $f: J_2(M) \to M$ by (1) $f(\{x\}) = x$ for all single-element sets; (2) $f(\{x,y\}) = x$ if x is in $[0, 1]$ and y is in $[2, 3]$; (3) $f(\{x, y\}) = x$ if x and y are both in $[0, 1]$ and $x < y$; (4) $f(\{x, y\}) = y$ if x and y are both in $[2, 3]$ and $x < y$. Clearly, if we had more components, we could have more such "flip-flops".

We can prove the following.

Theorem 2. *If M is a locally compact separable metric space, and there is a continuous selection $f: F_2(M) \to M$, then M is homeomorphic to a subset of a line.*

The requirement of separability is necessary, of course, because, e.g., an uncountable discrete space satisfies the remainder of the hypotheses, but cannot be imbedded in a line. We are not sure to what extent the requirement of local compactness can be relaxed.

Of course, the converse is also true.

I will give here the proof for M assumed compact, which shows the nature of the argument.

In that event, each non-degenerate component K is a continuum, and f induces a selection $f|F_2(K)$. By Theorem 1 we can conclude that K is an arc. From the Lemma, we conclude that only the end points of K can be limit points of $M - K$. By the Moore-Kline-Miller theorem [8], this is sufficient to show that M is homeomorphic to a closed subset of the interval. The fact that we assume only local compactness introduces technical complications without, however, changing the essential idea of the proof.

After completion of our joint paper, I began wondering about the possibility of generalization to non-metric spaces. Although we made no speical mention of it, the second proof of Theorem 1 in our joint paper (or Michael's methods in [7]) establishes that if H is a Hausdorff continuum, and there is a continuous selection on 2^H, then H is a "generalized arc" in the sense of my paper [10] or a "generalized 1-cell" in the sense of R. L. Wilder [9]. Other names have been used, but in essence H has almost all the properties of an interval that do not depend on separability.

It seems natural to ask next what one can say about selections for compact totally disconnected Hausdorff spaces. If such a space were metric, the situation is clear. Every such space can be imbedded in the interval and so continuous selections always exist.

For the case of the Hausdorff space, however, the problems are yet unresolved. I have proved the following result.

Theorem 3. *If H is a compact totally disconnected Hausdorff space, and $f: F_2(H) \to H$ is a continuous selection, then H has an order such that the order topology is the original topology for H.*

That is, H can be linearly ordered.

One's first thought is to proceed by defining $x < y$ to mean $f(\{x, y\}) = x$. Indeed, Michael used exactly this to define an order in connected spaces having a continuous selection. The trouble with this definition for non-connected spaces is that the relation need not be transitive, even for finite spaces. Nevertheless this is the basis for a construction by transfinite induction, which is conceptually simple, but which has an induction hypothesis of ten statements, some rather complicated, and I will not give the argument here.

Instead, I will discuss what was to me an unexpected connection with the representation of Banach spaces.

There is a classical string of results due to Banach and Mazur [1]. Consider a separable Banach space B. There is a compact metric space M such that B is isomorphic to a subspace of C(M), the Banach space of continuous real functions on M. There is a continuous function φ mapping the middle-third Cantor set K onto M. Functional composition lifts isomorphically C(M) onto a subspace of C(K); that is, each element g of C(M) is mapped onto the element $g\varphi$ of C(K). If f is an element of C(K), one can extend f to a real continuous function of f* of the same norm by defining f* linearly in the complementary intervals of K so that it agrees with f at the endpoints. Thus one finally has that <u>every separable Banach space is isomorphic to a subspace of</u> $C(I^1)$.

The particular compact metric space they chose was the very natural one of the linear functionals of norm ≤ 1, and one can define the desired continuous functions as evaluations.

Suppose that B is not separable. What parts of this program can one carry out? One can represent B as a subspace of C(H), H a compact Hausdorff space. Then one can find a compact totally disconnected Hausdorff space K that can be mapped onto H, and so lift C(H) into C(K). One such space would be the Stone-Cech compactification of the discrete topology of H.

Now if every compact totally disconnected Hausdorff space K could be linearly ordered, one could insert ordinary open intervals between each two consecutive points of K, and obtain a generalized arc I. By the same extension process, one could then represent C(K) as a subspace of C(I). This would be a most satisfactory generalization. I had gotten this far and had tried vainly to prove such orderings were possible when a surprising and elegant paper by Mardešić and Papić [6] appeared. This paper showed that the Hahn-Mazurkiewicz Theorem, that every locally connected metric continuum is the continuous image of the interval, had no generalization to locally connected Hausdorff continua. It was very easy to use their theorem to show that <u>there are totally disconnected Hausdorff spaces that cannot be linearly ordered</u> [11].

With Theorem 3, this result shows that there exist totally disconnected compact Hausdorff spaces that have no continuous selection. In fact, the existence of a continuous selector is necessary and sifficient for a linear order. We are left with a number of questions.

(1) Which compact totally disconnected Hausdorff spaces have a continuous selection (have a linear order)? Is there a reasonable subclass with continuous selections?

(2) In the Banach-Mazur construction, or its attempted generalization, the selection of the compact space at the first stage was done for convenience at that point of the construction. Is it possible in the non-separable case to define the compact (totally disconnected) space so that it can be ordered? For what non-separable Banach spaces is this possible? Can a selection be defined on the appropriate space by means of operations in the Banach space?

To revert to the problems of metric spaces, we can ask a question modified from one in Michael's paper [7].

(3) Suppose M is a metric continuum. Every collection of disjoint closed sets filling up M is a subspace of 2^M. Suppose that for each such subspace G there is a continuous selection $f: G \to M$. Is M necessarily a dendrite?

The difficulty with this question is that any one such subspace G may have no useful properties. Consider in our first example the decomposition G of X whose elements are the interval I, the intervals I_n, $n = 1, 2, 3, \cdots$, and the individual points of $I_0 - I$. The subspace G consists of a countable closed discrete set--the points $I, I_1, I_2, \ldots$ -- and a half-open interval. Any selection whatever on G is continuous in the subspace topology in 2^X for G. The supposition that M is a continuum does not prevent such subspace. Michael pointed out that M could not contain a simple closed curve. Hence this question reduces to asking whether M is necessarily locally connected.

Bibliography

[1] S. Banach, Theorie des operations lineaires, Warsaw, 1932.

[2] R. Engelking, R. W. Heath, and E. Michael, Topological well-ordering and continuous selections, Inventiones Math., 6 (1968), pp. 150-158.

[3] J. G. Hocking and G. S. Young, Topology, Addison-Wesley Publishing Co., Reading, Mass., 1961.

[4] K. Kuratowski, Topology, vol. II, Acad. Press and P. W. N., 1969.

[5] K. Kuratowski, S. B. Nadler, and G. S. Young, Continuous selections on locally compact separable metric spaces, to appear in Fundamenta Mathematica.

[6] S. Mardešić and P. Papić, Glasnik Mat. Fiz. Astr. 15 (1960).

[7] E. Michael, Topologies on spaces of subsets, Trans. Amer. Math. Soc., 71 (1951), pp. 152-182.

[8] E. W. Miller, On subsets of a continuous curve which is on an arc of the continuous curve, American Journal of Math., 54 (1932), pp. 397-418.

[9] R. L. Wilder, Topology of Manifolds, Amer. Math. Soc. Colloquium Publications, vol. 32, 1949.

[10] G. S. Young, The introduction of local connectivity by change of topology, Amer. Jour. Math. 68 (1946), pp. 479-494.

[11] ___________, Representations of Banach spaces, Proc. Amer. Math. Soc. 13 (1962), pp. 667-668.

Bitte wenden / Continued

Vol. 72: The Syntax and Semantics of Infinitary Languages. Edited by J. Barwise. IV, 268 pages. 1968. DM 18,– / $ 5.00

Vol. 73: P. E. Conner, Lectures on the Action of a Finite Group. IV, 123 pages. 1968. DM 10,– / $ 2.80

Vol. 74: A. Fröhlich, Formal Groups. IV, 140 pages. 1968. DM 12,– / $ 3.30

Vol. 75: G. Lumer, Algèbres de fonctions et espaces de Hardy. VI, 80 pages. 1968. DM 8,– / $ 2.20

Vol. 76: R. G. Swan, Algebraic K-Theory. IV, 262 pages. 1968. DM 18,– / $ 5.00

Vol. 77: P.-A. Meyer, Processus de Markov: la frontière de Martin. IV, 123 pages. 1968. DM 10,– / $ 2.80

Vol. 78: H. Herrlich, Topologische Reflexionen und Coreflexionen. XVI, 166 Seiten. 1968. DM 12,– / $ 3.30

Vol. 79: A. Grothendieck, Catégories Cofibrées Additives et Complexe Cotangent Relatif. IV, 167 pages. 1968. DM 12,– / $ 3.30

Vol. 80: Seminar on Triples and Categorical Homology Theory. Edited by B. Eckmann. IV, 398 pages. 1969. DM 20,– / $ 5.50

Vol. 81: J.-P. Eckmann et M. Guenin, Méthodes Algébriques en Mécanique Statistique. VI, 131 pages. 1969. DM 12,– / $ 3.30

Vol. 82: J. Wloka, Grundräume und verallgemeinerte Funktionen. VIII, 131 Seiten. 1969. DM 12,– / $ 3.30

Vol. 83: O. Zariski, An Introduction to the Theory of Algebraic Surfaces. IV, 100 pages. 1969. DM 8,– / $ 2.20

Vol. 84: H. Lüneburg, Transitive Erweiterungen endlicher Permutationsgruppen. IV, 119 Seiten. 1969. DM 10,– / $ 2.80

Vol. 85: P. Cartier et D. Foata, Problèmes combinatoires de commutation et réarrangements. IV, 88 pages. 1969. DM 8,– / $ 2.20

Vol. 86: Category Theory, Homology Theory and their Applications I. Edited by P. Hilton. VI, 216 pages. 1969. DM 16,– / $ 4.40

Vol. 87: M. Tierney, Categorical Constructions in Stable Homotopy Theory. IV, 65 pages. 1969. DM 6,– / $ 1.70

Vol. 88: Séminaire de Probabilités III. IV, 229 pages. 1969. DM 18,– / $ 5.00

Vol. 89: Probability and Information Theory. Edited by M. Behara, K. Krickeberg and J. Wolfowitz. IV, 256 pages. 1969. DM 18,– / $ 5.00

Vol. 90: N. P. Bhatia and O. Hajek, Local Semi-Dynamical Systems. II, 157 pages. 1969. DM 14,– / $ 3.90

Vol. 91: N. N. Janenko, Die Zwischenschrittmethode zur Lösung mehrdimensionaler Probleme der mathematischen Physik. VIII, 194 Seiten. 1969. DM 16,80 / $ 4.70

Vol. 92: Category Theory, Homology Theory and their Applications II. Edited by P. Hilton. V, 308 pages. 1969. DM 20,– / $ 5.50

Vol. 93: K. R. Parthasarathy, Multipliers on Locally Compact Groups. III, 54 pages. 1969. DM 5,60 / $ 1.60

Vol. 94: M. Machover and J. Hirschfeld, Lectures on Non-Standard Analysis. VI, 79 pages. 1969. DM 6,– / $ 1.70

Vol. 95: A. S. Troelstra, Principles of Intuitionism. II, 111 pages. 1969. DM 10,– / $ 2.80

Vol. 96: H.-B. Brinkmann und D. Puppe, Abelsche und exakte Kategorien, Korrespondenzen. V, 141 Seiten. 1969. DM 10,– / $ 2.80

Vol. 97: S. O. Chase and M. E. Sweedler, Hopf Algebras and Galois theory. II, 133 pages. 1969. DM 10,– / $ 2.80

Vol. 98: M. Heins, Hardy Classes on Riemann Surfaces. III, 106 pages. 1969. DM 10,– / $ 2.80

Vol. 99: Category Theory, Homology Theory and their Applications III. Edited by P. Hilton. IV, 489 pages. 1969. DM 24,– / $ 6.60

Vol. 100: M. Artin and B. Mazur, Etale Homotopy. II, 196 Seiten. 1969. DM 12,– / $ 3.30

Vol. 101: G. P. Szegö et G. Treccani, Semigruppi di Trasformazioni Multivoche. VI, 177 pages. 1969. DM 14,– / $ 3.90

Vol. 102: F. Stummel, Rand- und Eigenwertaufgaben in Sobolewschen Räumen. VIII, 386 Seiten. 1969. DM 20,– / $ 5.50

Vol. 103: Lectures in Modern Analysis and Applications I. Edited by C. T. Taam. VII, 162 pages. 1969. DM 12,– / $ 3.30

Vol. 104: G. H. Pimbley, Jr., Eigenfunction Branches of Nonlinear Operators and their Bifurcations. II, 128 pages. 1969. DM 10,– / $ 2.80

Vol. 105: R. Larsen, The Multiplier Problem. VII, 284 pages. 1969. DM 18,– / $ 5.00

Vol. 106: Reports of the Midwest Category Seminar III. Edited by S. Mac Lane. III, 247 pages. 1969. DM 16,– / $ 4.40

Vol. 107: A. Peyerimhoff, Lectures on Summability. III, 111 pages. 1969. DM 8,– / $ 2.20

Vol. 108: Algebraic K-Theory and its Geometric Applications. Edited by R. M.F. Moss and C. B. Thomas. IV, 86 pages. 1969. DM 6,– / $ 1.70

Vol. 109: Conference on the Numerical Solution of Differential Equ tions. Edited by J. Ll. Morris. VI, 275 pages. 1969. DM 18,– / $ 5.

Vol. 110: The Many Facets of Graph Theory. Edited by G. Chartra and S. F. Kapoor. VIII, 290 pages. 1969. DM 18,– / $ 5.00

Vol. 111: K. H. Mayer, Relationen zwischen charakteristischen Zahle III, 99 Seiten. 1969. DM 8,– / $ 2.20

Vol. 112: Colloquium on Methods of Optimization. Edited N. N. Moiseev. IV, 293 pages. 1970. DM 18,– / $ 5.00

Vol. 113: R. Wille, Kongruenzklassengeometrien. III, 99 Seiten. 197 DM 8,– / $ 2.20

Vol. 114: H. Jacquet and R. P. Langlands, Automorphic Forms on GL (VII, 548 pages. 1970. DM 24,– / $ 6.60

Vol. 115: K. H. Roggenkamp and V. Huber-Dyson, Lattices over Order XIX, 290 pages. 1970. DM 18,– / $ 5.00

Vol. 116: Sèminaire Pierre Lelong (Analyse) Année 1969. IV, 195 pag 1970. DM 14,– / $ 3.90

Vol. 117: Y. Meyer, Nombres de Pisot, Nombres de Salem et Analy Harmonique. 63 pages. 1970. DM 6,– / $ 1.70

Vol. 118: Proceedings of the 15th Scandinavian Congress, Oslo 19 Edited by K. E. Aubert and W. Ljunggren. IV, 162 pages. 1970. DM 12,– / $ 3.30

Vol. 119: M. Raynaud, Faisceaux amples sur les schémas en group et les espaces homogènes. III, 219 pages. 1970. DM 14,– / $ 3.90

Vol. 120: D. Siefkes, Büchi's Monadic Second Order Success Arithmetic. XII, 130 Seiten. 1970. DM 12,– / $ 3.30

Vol. 121: H. S. Bear, Lectures on Gleason Parts. III, 47 pages. 19 DM 6,–/$ 1.70

Vol. 122: H. Zieschang, E. Vogt und H.-D. Coldewey, Flächen u ebene diskontinuierliche Gruppen. VIII, 203 Seiten. 1970. DM 16,– $ 4.40

Vol. 123: A. V. Jategaonkar, Left Principal Ideal Rings. VI, 145 pag 1970. DM 12,– / $ 3.30

Vol. 124: Sèminare de Probabilités IV. Edited by P. A. Meyer. IV, 2 pages. 1970. DM 20,– / $ 5.50

Vol. 125: Symposium on Automatic Demonstration. V, 310 pages. 19 DM 20,– / $ 5.50

Vol. 126: P. Schapira, Théorie des Hyperfonctions. XI, 157 pages. 19 DM 14,– / $ 3.90

Vol. 127: I. Stewart, Lie Algebras. IV, 97 pages. 1970. DM 10,– / $ 2.

Vol. 128: M. Takesaki, Tomita's Theory of Modular Hilbert Algeb and its Applications. II, 123 pages. 1970. DM 10,– / $ 2.80

Vol. 129: K. H. Hofmann, The Duality of Compact Semigroups a C*- Bigebras. XII, 142 pages. 1970. DM 14,– / $ 3.90

Vol. 130: F. Lorenz, Quadratische Formen über Körpern. II, 77 Sei 1970. DM 8,– / $ 2.20

Vol. 131: A Borel et al., Seminar on Algebraic Groups and Relat Finite Groups. VII, 321 pages. 1970. DM 22,– / $ 6.10

Vol. 132: Symposium on Optimization. III, 348 pages. 1970. DM 22,– $ 6.10

Vol. 133: F. Topsøe, Topology and Measure. XIV, 79 pages. 197 DM 8,– / $ 2.20

Vol. 134: L. Smith, Lectures on the Eilenberg-Moore Spectral Sequen VII, 142 pages. 1970. DM 14,– / $ 3.90

Vol. 135: W. Stoll, Value Distribution of Holomorphic Maps into Co pact Complex Manifolds. II, 267 pages. 1970. DM 18,– / $

Vol. 136: M. Karoubi et al., Séminaire Heidelberg-Saarbrücken-St buorg sur la K-Théorie. IV, 264 pages. 1970. DM 18,– / $ 5.00

Vol. 137: Reports of the Midwest Category Seminar IV. Edited S. MacLane. III, 139 pages. 1970. DM 12,– / $ 3.30

Vol. 138: D. Foata et M. Schützenberger, Théorie Géométrique c Polynômes Eulériens. V, 94 pages. 1970. DM 10,– / $ 2.80

Vol. 139: A. Badrikian, Séminaire sur les Fonctions Aléatoires Li aires et les Mesures Cylindriques. VII, 221 pages. 1970. DM 18,– $ 5.00

Vol. 140: Lectures in Modern Analysis and Applications II. Edited C. T. Taam. VI, 119 pages. 1970. DM 10,– / $ 2.80

Vol. 141: G. Jameson, Ordered Linear Spaces. XV, 194 pages. 197 DM 16,– / $ 4.40

Vol. 142: K. W. Roggenkamp, Lattices over Orders II. V, 388 pag 1970. DM 22,– / $ 6.10

Vol. 143: K. W. Gruenberg, Cohomological Topics in Group Theo XIV, 275 pages. 1970. DM 20,– / $ 5.50